"A scientist takes on the study of passion . . . Modern medicine requires a delicate balance of scientific knowledge and clinical wisdom. A deep sense of humanity does not hurt at all. Anxiety and pleasure are terribly strong counterpoints in the fugue of life. Liebowitz senses how delicately it must be composed." —LOS ANGELES HERALD-EXAMINER

"A new, more sensible approach to psychiatry, bringing together the power of chemistry and the insights of self-understanding." —Alvin Toffler

"The biochemical nuts and bolts, so to speak, of romantic love . . . Combining some of the happier characteristics of both poet and scientist, he writes with insight, humor, and a genius for handling heavy scientific data with a light touch." —DENVER POST

Dr. Michael R. Liebowitz *is currently Director of the Anxiety Disorders Clinic at the New York State Psychiatric Institute, Assistant Professor of Clinical Psychiatry at the College of Physicians and Surgeons at Columbia University, and Assistant Attending Psychiatrist at Columbia-Presbyterian Medical Center.*

THE CHEMISTRY OF LOVE

Michael R. Liebowitz, M.D.

BERKLEY BOOKS, NEW YORK

This Berkley book contains the complete
text of the original hardcover edition.

THE CHEMISTRY OF LOVE

A Berkley Book / published by arrangement with
Little, Brown and Company

PRINTING HISTORY
Little, Brown edition published 1983
Berkley edition / June 1984

ISBN: 0-425-06989-3

A BERKLEY BOOK ® TM 757,375
Berkley Books are published by The Berkley Publishing Group,
200 Madison Avenue, New York, New York 10016.
The name "BERKLEY" and the stylized "B" with design
are trademarks belonging to Berkley Publishing Corporation.

PRINTED IN THE UNITED STATES OF AMERICA

To Sam, who made me want to write
and
To Sharon, who gave me something to write about.

ACKNOWLEDGMENTS

Many people helped bring this book into being. I am grateful for what I learned from my teachers, most especially Donald F. Klein, M.D., from my colleagues, and from my patients. Family and friends provided encouragement throughout, as did my agent Charlotte Sheedy and her staff. Hillard Kaplan helped with the cross-cultural research. My secretary, Inez Eveline Elliot, helped type the manuscript; so did my wife, Sharon, who also did much more. Donald F. Klein, Norman Rosenthal, and Henry Geller critiqued early drafts. My editor at Little, Brown, Barry Lippman, was a pleasure to work with and played a great role in shaping the book.

One further point. The case histories are based on real patients, but all identifying details are fictional.

MRL

CONTENTS

THE CHEMISTRY OF LOVE

INTRODUCTION

Why do some people take it so hard when a romance ends? Why do others put up with so much abuse from a partner or act so submissively to keep things from ending? Why do some find it intolerable to be alone, or feel desperate when not in a romantic relationship? And why do some people persist in choosing partners who can't or won't come through for them?

Traditionally such questions were answered in psychological terms. A person's response to rejection reflected his or her self-esteem. People who value themselves highly can be hurt by romantic rejection but not devastated. They put up with feeling mistreated in a relationship for only so long. Others, with a lower sense of self-worth, experience rejection more deeply and tolerate greater abuse from their partners rather than call an end to things.

In the same vein psychologists often view a person's discomfort over being alone as reflecting a lack of maturity. People who have become independent can tolerate being alone or unattached and still enjoy their lives; others, who have not outgrown dependency, can't. How we choose partners is viewed in a similar way — people who are desperate for, or ambivalent about, relationships will choose more hastily or unwisely than those who are not.

This way of seeing our behavior, while correct insofar as it goes, ignores a crucial aspect of why we act as we do. In all areas of human endeavor, and most especially in a highly charged area like romance, we are governed by biological as well as by psychological processes, by our brains as well as our minds.

Most of us have grown up in a psychologically sophisticated culture that pays a great deal of attention to how our minds work, and how, through one form or another of psychotherapy, we can change if we so desire. But what has been given far less attention is how our thoughts and feelings are also shaped by our biology — by the workings of our brains.

It may be startling to realize that every thought and every feeling we have, every action we undertake occurs only because of some form of biochemical activity in our brains — some electrical action along a nerve pathway, some chemical flux across the spaces between nerve cells. If we meet someone whom we find particularly attractive, and they seem to respond in a similar way, most of us feel excited and happy, even elated, as well as more optimistic and energetic. It's as though someone slipped us a stimulant. If we meet the same person and they are disinterested or rejecting, our body processes shift in the opposite way — we feel sadness and self-doubt, accompanied by pessimistic thoughts and reduced energy. In this case it's as though the natural stimulant flow in our bodies was suddenly shut off.

There is growing evidence that the differences in how people react to positive and negative romantic experiences are determined by differences in the way they respond biologically to these situations. Specifically, the intensity of one's excitement on meeting someone new, how long the "honeymoon" stage of a romance lasts, or the depth and duration of a post-romance depression, are ultimately determined by the changes that occur in our brain chemistry and how long those changes persist. While whom we fall in love with is undoubtedly influenced by psychological factors, none of us would have any romantic experiences — no thrills, no heartbreaks, no drive for enduring relationships, were it not for the way we as a species are biologically wired.

My own thinking has gone through a three-stage process, which I believe has great implications for how we think about human emotions. The first step was my discovery that adding a biological perspective helped me to become enormously more effective in understanding and working with many individuals experiencing romantic difficulties.

In concrete terms what this means is that many people who suffer crushing depression when a romance ends, or who are so dependent on relationships that they cannot enjoy their lives when not in one or assert themselves when they are, or who become so giddy and elated with romantic attention that they continually get involved with the wrong people, or who are so sensitive to even small slights that they are continually feeling hurt or mistreated by the people they love, seem to be suffering from problems that are at least in part biologically based, and can be enormously helped by treatment with currently available psychiatric medications.

The second stage developed as I started to ask myself: Is love itself a biological process? That is, if extreme or disabling romantic highs and lows have a biochemical basis, could this in fact also be true for the more "normal" ups and downs all of us experience depending on how our love lives are going? Looking into the matter more closely I began to realize that no one had ever seriously tried to examine the biochemical basis for our romantic drives. The fact is, if you want to look seriously at why we pursue romance, or why we feel high when we fall in love, or secure but also restless at times when a romance is well established, or uneasy when separated from a loved one, or devastated when a relationship ends, you have to ask what biochemical processes are going on in the human brain at such times.

The third stage came when I began to notice many similarities between romantic highs and lows and other intense feelings. If brain chemistry does play an important role in romantic drives and reactions, could what goes on in our brains at such times be similar to what occurs during other emotional states? The answer seems to be yes. As we will see, romantic ups and downs are very closely related to other profound highs and lows we experience, be they the joys of discovery, success, or beauty, or the threats or hurts of other losses or disappointments. The bottom line seems to be that our capacities to experience any joy or excitement, without which we would not want to continue living, or to feel hurt, sad, or scared, without which we would not be human, all reside in a delicate biochemical balance of our nervous systems. When this balance shifts in certain ways, our reaction is to feel excitement or pleasure. Shift the chemical balance in another direction, however, and the opposite emotions ensue.

Relating psychological and emotional experience to biochemical changes and vulnerabilities is something very new. While biochemists, physiologists and pharmacologists, among others, have for many years studied how our brains work, there has been little attempt to apply these findings to understanding our emotional reactions to romance or other important human pursuits. Instead, this has been left to schools of thought that tend to focus exclusively on how prior life experiences and/or social forces shape the way we perceive, interact, and react to life.

This tendency to separate biology from our important emotional experiences has led to a kind of psychiatric either/or thinking. This started becoming apparent to me in the first weeks of my psychiatric training, when I was assigned to a short-term in-patient ward. One of my first patients was a thirty-year-old woman who was admitted after taking an overdose of pills in what was more a gesture than a serious suicide attempt. The immediate stress involved her boyfriend, who wanted to break off with her. Severe depression resulted from this rebuff. However, it turned out that several of her previous relationships had ended in a similar way.

What I remember best about this patient was that several of my supervisors disagreed so strongly about what was going on. One, with sophisticated psychoanalytic background, said: "Your patient has a personality structure that is still quite infantile and long years of psychotherapy will be required to help her become more mature." Another, who tended to rely more on medication, said: "Your patient is highly vulnerable to a certain type of depression and a particular antidepressant is what is required." What seems true in retrospect, although none of us understood it at the time, was that this patient, and many like her, had *both* biological and psychological difficulties, and *both* needed to be addressed.

Perhaps because of experiences such as this I became interested in those kinds of problems where both biochemical and psychological factors had to be considered. As a result, I am often asked to consult in situations where attention to only one or the other dimension has not been productive. In many such cases attention to the neglected aspect of things, whether it be psychological problems in a patient treated up to that time only with drugs, or to biochemical disturbances in a patient previously treated only with psychotherapy, has been extremely helpful.

It first became apparent to me a number of years ago that biological vulnerabilities play an important but unrecognized

The Chemistry of Love

role in how some people feel about and function in romantic relationships. While treating patients with one or another type of antidepressant medication, I saw some startling changes. Initially, their moods improved and their ability to function returned. But that's not so surprising — after all, that's what we expect from such medication. But I also began to observe something else — shifts in attitudes, increases in self-esteem and independence, decreased reliance on others — changes that one did not expect to occur from taking medication.

A patient from whom I learned a great deal comes to mind. Several years ago I was asked to see a young married woman who was in a painful depression. A month previously her husband had moved out and begun an affair with another woman, leaving her to manage the house and her own full-time job. Worse yet, he now wanted to move back in and be able to entertain his new woman friend in the house while my patient was not at home. She felt that this last demand was outrageous, but was fearful of telling him so because he might decide to seek a divorce.

Now I'm sure that many of you are saying, "What has she got to lose?" or "Why is she being such a namby-pamby about it? Why not tell the guy that he is acting like a selfish jerk and that he should shape up or ship out?" I must confess that inwardly this was my own reaction. But further delving revealed that this lady had suffered several previous depressions in her life and was fearful that she'd wind up feeling much worse if she and her husband broke off completely. Thus she was in a quandary — the current situation was extremely painful, but standing up for herself was scary because it might, in her opinion, make things worse.

Well, we have to help this lady become more independent, right? The only problem was that although this was the first time I was seeing her, she'd been in treatment at this particular clinic for several years; the main focus of the treatment had been to help her become more independent of and confronting with her ne'er-do-well husband. All of which had led him to say, "I don't like you acting like this," then move out and take up with his new girlfriend. Thus my patient in some ways felt that her treatment at the clinic had made things worse, and that she would have been better off being more submissive to her husband.

My first goal was to treat her depression, so I started her on antidepressants and saw her weekly while gradually building up the dose. I also advised her not to do anything definitive with regard to her husband until she was feeling a little better and could think more calmly about the matter.

A funny thing happened. Within about three weeks her mood, energy, appetite, and ability to sleep all began to improve, which was expected. But also, her outlook on the problem with her husband began to change, much to the surprise of the clinic staff that had been working with her for several years. One week she came back and said, "You know, this thing with my husband is ridiculous. I told him he has to choose. Either he comes back to live with me and is faithful, or he stays with his new girlfriend and we get a divorce." What had happened is that she no longer desperately feared being alone. She still wanted her husband back, but only if he would get his act together and behave like a husband. If not, she was prepared to get a divorce, manage on her own, and, if things worked out, find a new partner who would treat her in a better way. This was a curious phenomenon, for antidepressant treatment seemed to improve not only a temporary low mood state but also long-standing problems in self-esteem and independence.

At first it was rather startling and ran counter to the traditional teaching and belief that these are exclusively "psychological issues." What I have been forced to conclude, however, after seeing numerous examples such as this, is that our abilities to deal with rejection by or separation from people important to us, to feel okay when alone or not in a relationship, to exercise good judgment in choosing romantic partners — all of which are crucial to personal happiness and successful relationships — seem to depend on our biological as well as psychological makeup. Moreover, many people who have difficulties in these areas appear to have largely unrecognized biochemical vulnerabilities and can be dramatically helped, first by recognition of this fact, and second, by treatment with currently available psychiatric medications, which help to normalize and stabilize their body chemistry.

This book will be about these patterns of sensitivity to rejection, low self-esteem, craving for romance, dependence on relationships, and especially their previously unrecognized biological aspects. Ongoing research is opening a whole new way for understanding and helping many people with serious interpersonal difficulties. It is also opening a whole new dimension for understanding normal human responses in romantic and other relationships. The highs we feel when falling in love or otherwise receiving approval, and the pain we suffer when feeling rejected, abandoned or isolated, are in part due to changes in brain activity and body chemistry that are triggered by what we are experiencing. What may distinguish most people from those who are overwhelmed by certain types of romantic or other emotional difficulties is that ultimately their *biological* reactions are not so severe or so long-lasting.

Let me give another example. Anxiety reactions, both extreme and normal, seem to play a role in our responses to separation. Extreme anxiety reactions to separation often lead to school phobias in children and panic attacks or crippling emotional dependencies in adults. For years such difficulties were treated, usually unsuccessfully, with long-term intensive psychotherapy. Recently, however, we have found that certain school phobias in children and panic reactions in adults can be dramatically helped by the same medications that block or reverse depressions, even if the people involved are not suffering from depression. It now looks as if children with school phobias and adults with panic attacks may be suffering from a common faulty brain alarm mechanism. Moreover, people who have suffered for years can often be helped in a matter of weeks to months.

The importance of this for our discussion of romance is that problems in this same brain alarm system may cause excessive fears of being alone (high separation anxiety). This can really make a mess of your love life, because it will cause you to jump into relationships too hastily or to stick too long with a partner who mistreats you or with whom you are not happy. Moreover, more muted problems of this same brain alarm system could (this is speculative) manifest itself in such tendencies as feeling anxious when alone, compulsively needing to socialize, or a high vulnerability to feeling lonely.

It is not my intention to view human behavior in exclusively biochemical terms. This would be as unbalanced as looking only at psychological or social factors. We human beings are complex organisms that are shaped by biological, psychological, and social forces. What I am trying to do is restore a greater balance to our view of human behavior and emotional experience, which has been dominated for the past several decades by psychological and sociological ways of thinking. Granted, many biologists have seen us as shaped by evolution and have drawn parallels between animal and human mating, courtship, and bonding patterns (suggesting our emotional reactions are influenced by inherited tendencies). However, our discussion really begins where others leave off, since even if some behavior pattern or emotional tendency is inherited, this tells us nothing about what actually goes on within us to make us act, react, or feel as we do. To understand this, we have to look at our chemistry.

When I talk about treating emotional difficulties with psychiatric medications, a number of objections may come to mind, such as: "He's doping people up"; "He is simply numbing them"; "He's destroying their ability to feel"; "He is treating symptoms, not causes"; "He's creating an artificial sense of euphoria"; "He is causing damage by giving drugs with unknown side effects."

I would like to address some of these issues at the outset. With regard to treatment, the antidepressants have helped an untold number of people to *achieve* or *regain* satisfying levels of functioning with few or no adverse consequences. Moreover, many of these people had tried years of psychotherapeutic treatment without achieving the desired results. I myself am a trained therapist who has and continues to work psychotherapeutically with many patients. But one of the great advances of modern psychiatry is the recognition that different conditions require different treatments. That sounds obvious, but in fact it's not. Some emotional problems benefit from psychotherapy alone, some from pharmacotherapy (drugs) alone, and some from a combination of the two.

Moreover, the types of psychiatric drugs I will be discussing do not create artificial states of euphoria or numb normal human feelings. Rather, they help by diminishing excessive reactions of depression, anxiety, rejection sensitivity, inappropriate elation or irritability. Many psychiatric disorders involve states of altered brain activity that cause a loss of normal emotional control, leading to the experience of extreme feelings (anxiety, depression, euphoria, rage) unwarranted by actual events. Antidepressants and antimanic drugs work not by creating artificial moods, but by restoring more normal emotional control. For this reason they tend not to be abused by the general population, since they have little effect (except, in some cases, sedation) in people whose emotional regulatory mechanisms are not out of line to begin with. This is also true for the antipsychotic drugs like Thorazine, Stelazine, Haldol, etc., which can, however, make people look zombielike if certain side effects are neglected or if they are given in doses that are too large, or if they are prescribed for people who don't really need them. Some patients (and most likely some readers as well) fear that taking antidepressants or lithium will make them look or feel like zombies, but as I've said, this does not occur with these types of medications, although it does at times with antipsychotic drugs.

One final caution. What I've said about psychiatric drugs does not apply to sedatives, stimulants, hallucinogens, narcotics, or marijuana, which can alter mood states in anyone who takes them, and therefore are widely abused.

Extreme sensitivity to rejection, excessive emotional dependency, and panic attacks can be *greatly diminished or blocked by specific medications that do not in any way impair or otherwise modify a person's ability to react emotionally.* If someone no longer reacted to rejection, but looked like a zombie and also could feel no joy or pleasure, then it would appear that we were simply numbing that person's capacity to feel anything. But if what we see instead is a newfound ability to withstand rejection or criticism, coupled with no loss of ability to feel anything else, then we can be more confident about having normalized things in a disturbed biological system.

In the end, I hope readers feel this book is about more than romance, because I also will try to share what we are learning about the biological aspects of elation, joy, depression, sadness, and anxiety. Along the way, there also will be discussions of panic, fears of separation, needs for attention and applause, and feelings of self-esteem, self-worth and self-confidence. There are many exciting things happening in this field today, and I hope you keep in mind throughout that I am not attempting to do away with traditional psychological explanations of feelings or behaviors, but rather, to enrich and amplify them with the new findings of brain chemistry and psychopharmacology.

CHAPTER 1

BRAIN CHEMISTRY:
"The Doors of Perception"

My first thoughts about the chemical workings of our minds actually did not occur in an entirely professional setting. Like many present-day psychiatrists I was interested in psychology while still in college. But that was the heyday of psychoanalysis and behaviorism, where both normal behavior and mental illness were seen largely in the light of one's prior experiences, both real and imagined. At that time, the recent revolutions in psychobiology and psychopharmacology were still a long way off.

Meanwhile, outside the classroom a form of psychobiology was sweeping the campuses — in the mid-1960s the drug scene began to flourish. All around me I saw the mind-altering effects of marijuana, hashish, and the psychedelics. It was a crash course in experimental psychopharmacology — how thought processes, feelings and experiences could be altered by chemicals.

I remember being particularly impressed by the reported effects of the psychedelics — LSD, psilocybin, and mescaline. The very basics of experience — our sense of time, space, and logic — seemed to be altered by these drugs, suggesting that the way we usually experienced things was largely a product of a particular state of brain chemistry. On LSD trips people reported that forms flowed into each other, time and space seemed less distinct, and logical contradictions sat side by side, both feeling perfectly meaningful and true. The human mind appeared to have strange capabilities not normally used. Our usual way of seeing the world around us seemed to be based on our normal chemical balance. Alter this chemical balance, as one did by ingesting LSD, and one began to perceive the universe in very different ways.

A key example of this is the so-called mystical experience. For centuries certain individuals have described brief, extraordinary experiences in which they feel transported out of their ordinary states of consciousness, and go on to experience a sense of merging with the universe, of being one with all of nature and all of history. At such times the physical boundaries of the body and the time limits of one's own life, of which we are normally so aware, fade into unimportance. Instead one feels oneself as a small part of the endless flow of time, the endless process of life.

Most often these states (or religious experiences) are described in ecstatic terms, both because of their novelty and the solutions they offer to worries about our own mortality. If I am simply a small chunk of the universe, a condensation of the universal ebb and flow that came together to form me and will simply flow into something else when my life ends, then dying is really nothing to fear. Similarly, most of the mundane things with which I normally concern myself will seem far less important.

Because of their uniqueness, beauty, aura of truthfulness, and powerful spiritual message, such mystical states have long been considered "higher truths," allowed to a few individuals because of holiness, divine inspiration, or ardent seeking. Imagine then when people began to report similar "mystical" experiences after taking psychedelic drugs. In *Doors of Perception*, Aldous Huxley described his own experiences with mescalin. The title of his book was deliberately taken from a poem by the eighteenth-century mystical poet William Blake, who wrote:

> If the doors of perception were cleansed,
> Everything would appear to man as infinite.
> For man has closed himself up
> Till he sees all things thro' narrow chinks of his cavern.

Blake was suggesting that our normal, everyday perception of the world is a pale reflection of what it is really like, i.e., that we are capable of much more, as reflected in the brief glimpses of "true reality" available to society's mystics.

What Huxley noted was the similarity of his own mescalin experiences to the mystical experiences described by others. Huxley was in fact saying that mescalin could open up the "doors of perception," by removing the "gauze" that normally covers our perceptive apparatus.

As a latent psychobiologist I was very impressed with the model of a perceptual filter. What this means scientifically is that our brains function so that we see and interpret the world in particular ways. Tamper with this filter, as seekers have been doing through the ages by prolonged fasts, strenuous climbs, endless meditation, or ingestion of mind-altering drugs, and reality changes. Or more accurately, our perception of reality changes, sometimes along the lines of the so-called mystical experience. To me this was not a higher reality, as some claimed, or crazy, as others would have it, but simply different and unusual. It had its pluses — e.g., a powerful spiritual feeling — and its minuses — feeling one with the universe seemed to make it more than a little hard for people to function on a day-to-day basis.

Other budding psychobiologists were similarly impressed. If transient alterations of our perceptual filter could lead to mystical states, might not different, or longer-lasting, alterations lead to a different set of mental states — the psychoses?

A psychosis, for the great majority of you lucky enough never to have experienced one, is a mental state characterized by a break with our everyday sense of reality. Some people hear voices talking to or about them when no one is actually speaking. Others become firmly convinced of some belief — of a conspiracy against them, or a cancer within them, or that they are Napoleon or Jesus — which cannot be shaken by any amount of contrary evidence. Thought processes may be scrambled, emotions blunted or wildly intensified. Sometimes the state comes "out of the blue"; sometimes as a response to stress that a person can no longer handle. Psychoses can also be brought on by excessive use of drugs like cocaine or amphetamine.

In the first flush of research enthusiasm the psychedelic drugs (relabeled "psychotomimetics") were thought to induce psychosislike states. In fact, a few poor souls from among the many who took LSD did wind up in the Bellevue emergency room. For the most part, however, those who took these drugs did not experience actual psychoses. Often someone taking LSD would hear voices when no one was talking, as would someone with a schizophrenic psychosis. But while the schizophrenic would actually believe that the voices were coming from somewhere outside his own head, the person on LSD usually understood that the drug he had taken was causing him to hear voices. One's touch with reality was not lost the way it is in a true psychosis.

This distinction notwithstanding, the psychedelic experience was a powerful demonstration of how our mental functioning depends on brain biochemistry, and how altering this chemistry can lead to a variety of unusual emotional experiences. This in turn provided a powerful stimulus to study brain function and its relation to emotional states.

* * *

There were two other developments in the 1960s that also stimulated great interest in the chemical functioning of our brain. Unlike the psychedelic experience, these took place within professional confines. I am referring here to the phenomenal developments in psychobiology and psychopharmacology that began in the mid-1960s.

Psychobiology is a field that focuses on the interrelationship of psychological and biological events, and by doing so, attempts to figure out the connections between mind and body. It is still a young science, but it has expanded rapidly in the last two decades. Let me give you an example from the field of depression.

That people became depressed or melancholic has been known for thousands of years. One can find sad, brooding, unresponsive figures in art and literature of many eras and cultures. But what caused this to happen was always a subject of speculation and conjecture. Inherited tendencies, peculiar temperaments, psychological conflicts, supernatural forces — all were invoked at one time or another.

It was also known for many years that the blood level of cortisol, a naturally occurring cortisonelike substance, goes up when human beings are under stress, presumably as part of the body's coping attempts. Since depressed people are under stress, and that stress is sometimes affected by psychotherapy, Dr. Edward J. Sachar decided to study cortisol fluctuations in depressed patients undergoing psychotherapy. He reasoned that changes in ways of coping brought about by psychotherapy should be accompanied by changes in blood cortisol levels.

Dr. Sachar, former Chairman of the Psychiatry Department at Columbia University and Director of the New York State Psychiatric Institute, never did finish his study of psychotherapy and cortisol. Rather, his attention was diverted by the striking finding that many depressed patients seemed to have cortisol levels that were elevated beyond what could be explained by the stress they were under. Moreover, their cortisol levels were much higher than patients suffering psychoses or anxiety reactions, who were under at least as much stress. Finally, the cortisol levels of the depressed patients differed in a peculiar way — the daily rhythm was thrown off. Instead of being high in the late morning and early afternoon and falling off in the evening and early morning, as is normal, the cortisol of a depressed patient stayed high all the time.

The importance of this finding was that it demonstrated a concrete biochemical abnormality accompanying depression. It's not that we think that the high cortisol causes the depression, but rather, that both the depressed mood and the high cortisol are caused by a derangement in brain chemistry. The abnormal cortisol rhythm is one example of this derangement, the low mood another. When a depressed person fully recovers, which may be some time after his mood begins to lift, his cortisol pattern also returns to normal.

This is the field of psychobiology — a search for irregularities in the body's functioning that accompany abnormal or heightened emotional states, as a way of understanding how things like depression come about. Now that we know that certain depressions and cortisol abnormalities go together, a crucial piece of the puzzle has been found. If we can now figure out how the brain regulates cortisol levels and what chemical processes go awry to cause the abnormal pattern seen in depression, then perhaps the causes of the mood difficulties will also be found.

Another great impetus to our understanding of mental and emotional functioning has been provided by recent developments in psychopharmacology. In a sense psychopharmacology is an old science — the mind-altering capacities of certain chemicals have been known for centuries. In the 1880s and 1890s dozens of patent medicines containing substances like opium and cocaine were widely used. In fact, the popular "soft" drink

Coca-Cola was in its early years a "hard" drink, since it contained cocaine, which was removed in 1903. Presumably Coca-Cola got its name from the coca leaf, which is the naturally occurring source of cocaine.

Barbiturates, the archetype of all "downs," were introduced into medical practice in 1903, while amphetamine, the prototypical stimulant or "upper," became available in the 1930s. From a psychiatric point of view these were nonspecific drugs that, while profoundly affecting brain processes, did not treat or reverse any specific mental states. It was not until the 1950s that more specific drugs became available in the form of antipsychotics and antidepressants. These latter drugs have revolutionized the practice of psychiatry.

The antipsychotics, while often suppressing rather than curing illness, have allowed many people to lead productive, independent lives where formerly they would have required continuous hospitalization. The antidepressants in fact also initially suppress symptoms, presumably by counteracting the underlying disturbance, so that many depressed patients will relapse if the drugs are discontinued too quickly. But in many depressions, and in some psychoses, the underlying biochemical disturbances are corrected in time; then the drugs can be stopped and normal functioning continues.

These drug effects also provide a powerful tool for investigating the way our brains work. Insofar as we know how an antipsychotic or antidepressant works biochemically, the fact that a given drug helps in one condition but not another tells us a great deal about the specific chemical disturbances of each of these conditions. If a widely used antidepressant causes an increase of brain chemical A but not brain chemical B, and helps depression but not schizophrenia, then the activity of chemical A is more likely decreased in depression than in schizophrenia. So drugs tell us something about the underlying biochemical abnormalities, but, as with psychedelics and psychobiology, they provide clues, not full solutions, to the puzzle of brain function.

I have used the term "puzzle" to speak of the brain, but perhaps mystery would be more apt, for when you stop to think about it for a moment the whole thing is pretty mysterious. As you read this book thoughts go through your mind; perhaps you get distracted for a moment and stare out a window thinking of something else. You may also have some feelings about the book, pleasurable I hope, or perhaps something else. But have you ever considered how all this happens? What goes on in your brain to make all this possible? What does it really mean to have an emotion? How do we come to have feelings? How does it all happen?

We all tend to speak freely of our emotions without stopping to think what they really are. Emotions are generally considered to have two elements — a feeling component and bodily arousal.

The feeling part has to do with whether something makes you happy, sad, angry, or frightened. Some experiences don't make you feel anything in particular — they are emotionally neutral.

Feelings are also accompanied by body sensations. For example, when you are afraid, you may notice such things as a cold sweat, a pounding in your chest, a sinking sensation in your gut. If you stop to consider, all feelings are to some degree associated with bodily sensations — people "jump" for joy because their legs feel springy and they feel energized. One also feels "weighed down" with sorrow or grief, lethargic or restless when bored, and tense or ready to explode when angered. Thus an emotion is a response to a person, object or situation in which you have both a physical bodily response and a feeling response.

The exact relationship between these physical and feeling aspects of emotions have been a matter of debate by psychologists for many years. Before the turn of the century William James proposed that emotions were based on our experience of our body sensations. That is, when we perceive a situation, James thought, our bodies react first, and our experience of this particular bodily reaction is the emotion. Thus, if we sense our body becoming tense in a given situation, we might then feel afraid; if it energizes us, we might then register excitement. This has come to be known as the James-Lange theory of emotions.

When I first read this I said to myself — Yes, there are times when things seem to happen this way, when our body's reaction leads us to realize that we are sexually aroused or angry, for example. I once had a boss who would take it out on his employees whenever he had a bad day. At such times he would rake us over the coals for some small error. I remember him yelling at me one day, and then looking down to find my fists clenched, making me realize that despite the contrite expression on my face I was furious.

Keep in mind that in some situations, our bodies are not as inhibited as our minds. As we have learned since William James, it's really not that our bodies react first, but rather, that we respond to our perceptions of events with many parts of our nervous system at the same time. In a situation like that with my autocratic boss, the angry feelings were temporarily blocked from my awareness by my embarrassment or fear, while my body, not feeling similarly constrained, reacted appropriately.

The problem with all this as a definition of what an "emotional response" is is that most of us have also had experiences in which we had a certain feeling but our bodies did not respond accordingly. Do these qualify as spontaneous emotional responses? The same inhibitions that sometimes block our awareness of feelings can also stop our bodies from reacting. Someone can be paralyzed by fear and not react to perceived danger, or be made impotent by guilt despite feeling very sexually aroused, etc. It seems that our feelings and our bodies can at times operate somewhat independently of each other in the area of emotions.

This is confirmed by situations in which physical arousal occurs but does not register at all as an emotional experience. Something may cause our blood pressure to go up or our stomach acid to flow freely, but arouse no accompanying feelings, and thus we remain emotionally if not totally oblivious to the event. In fact many people have to be taught to recognize this very thing — that is, to learn when their bodies are unduly tense, their blood pressure too high, or their pulse too rapid.

If the James-Lange theory were true, if emotion is the brain's awareness of our bodily reactions to events, then a variety of bodily reactions would have to exist to allow for different emotions. In fact for many years scientists searched for these different physiological reactions, but for the most part were unable to find them. What they discovered instead was that such diverse emotional experiences as anger and fear are accompanied by rather similar patterns of bodily arousal.

In time this led to the discovery of our sympathetic nervous system (SNS), which, when activated, prepares us for either flight or fighting. If we suddenly perceive danger, our SNS is stimulated and this in turn readies our bodies to meet the challenge. An internal alarm is sounded, and messages go out over the SNS network for our hearts to pump faster and harder, for our digestive system to slow down, for our sphincters to tighten up, for our blood sugar and adrenaline levels to rise, and for our air passages to dilate. All our resources are mobilized for maximum muscular effort, be it to fight whatever is threatening us if we think we can win or are cornered, or to run like hell if that seems more prudent and possible.

If the same physical arousal patterns occur with such different emotions, some scientists reasoned, then something besides bodily sensations must help determine what we ultimately feel. Specifically, they wanted to study what role our minds play in shaping our emotional reactions.

In the early 1960s two social psychologists, Drs. Stanley Schacter and Jerome Singer, carried out a set of experiments designed to examine the mind's contribution to emotional reactions. Student volunteers were asked to participate in a study of the effects of vitamin supplements on vision. Those who agreed were given an injection of the supposed vitamin, but in fact received either small doses of adrenaline or saltwater. Adrenaline is a drug whose effects are similar to stimulation of the SNS or fight-flight system described above. In the doses given by Schacter and Singer, the subjects who received adrenaline usually felt palpitations, tremor, and sometimes a feeling of flushing or breathing faster. Some subjects who received adrenaline were told that they might experience the symptoms that adrenaline actually produces, others were not told anything about possible drug effects, and still others were deliberately told that their feet might feel numb, they might feel itchy or could develop a slight headache, which are all reactions that adrenaline does not really produce. What the experimenters wanted to see was if giving people adrenaline as opposed to saltwater, and hence arousing them physically, would result in stronger emotional reactions to subsequent experiences. They also wanted to compare the reactions of people who got adrenaline but were informed about its effects with those who got the drug but were uninformed or misinformed. The hypothesis was that if when your heart pounded fifteen minutes after your shot and you knew it was a drug effect, your reaction would be different than if your heart was pounding and you did not attribute it to the drug. Specifically, the latter group might think what they were experiencing was making their hearts pound, and then rate their emotional reactions as stronger.

Finally, the researchers wanted to see which if any emotions adrenaline made stronger. Did it enhance only negative feelings, or would it amplify happy feelings as well? If adrenaline could make people feel both more happy and more angry, it would suggest that the underlying chemistry of these diverse emotions was the same. Some other factors besides our bodily arousal would then have to determine exactly what feeling we actually experienced in any given situation.

To test this, Schacter and Singer put the adrenaline- and saline-treated subjects into several supposedly emotional situations. Some were placed with a researcher posing as a second subject, who was trained to act euphorically while supposedly filling out research questionnaires. This person doodled on the forms, then crumpled them up and took basketball shots at the wastebasket. He also started flying paper airplanes, made slingshots out of rubber bands, built a tower out of research folders, and played with a hula hoop "left" in the room, all the time encouraging the research subject to participate. Other students were joined by a person, again acting as a stooge, who was trained to carry on in an angry fashion. This person complained about getting an injection, objected over and over again to various questions on the research forms, and finally ripped them up and stamped out of the room.

All subjects were observed through two-way mirrors, and were evaluated on their emotional reactions as well as asked to rate themselves on what they had experienced. Of the subjects who spent time with the "euphoric" stooge, and also received an injection of adrenaline, those who had been accurately appraised of possible adrenaline effects did not feel or act as happy as did those who were uninformed or misinformed. The researchers found that those who attributed their racing heart and sweaty palms to the drug were less likely to have their mood influenced by the stooge. Conversely, those who felt all stirred up inside and did not know why were more likely to get caught up in the stooge's mood and behavior.

For the students exposed to the mock-anger situation, the results were in some ways more striking. Those who received adrenaline and were uninformed about its possible effects were observed to behave more angrily than either the placebo group or those who were informed about drug effects. These results were considered to show that adrenaline enhanced angry feelings, and that feeling aroused internally, provided you didn't know why, also predisposed to greater anger.

The Schacter-Singer experiment had a great impact of modern psychology. Especially important was the conclusion that adrenaline could enhance both euphoric and angry feelings. This was taken to mean that internal bodily arousal contributed to emotional experience, but could enhance any feeling state that the mind considered appropriate. That is, adrenaline arousal would make you more happy if you were in a pleasant situation, more angry if in a frustrating one. Basically it seemed your mind chose the direction for your feelings, and your body reaction only gave the feelings their intensity.

In a large measure as a result of these experiments psychologists began to focus more on the mental aspects of emotional experience. In some ways this was quite sensible, for in everyday situations our minds interpret our experiences for us and tell us what to feel. In this sense, they act like vast computers. All new inputs are matched up with already existing data, which has been supplied both by experience and instinct. While instincts are harder to see in man than in animals, some human social reactions may be innate, such as an infant's smile response to a human face. In fact, children born without sight still smile, indicating it is not something we learn by imitation. However, most of our reactions are acquired through experience, which over time provides us with a vast array of pleasant and unpleasant memories. All new experiences or encounters are quickly checked by a scan of our internal "data banks." Those that resemble unpleasant past experiences will elicit negative feelings and those closer to previous pleasant experiences, positive ones. Neutral or seemingly novel stimuli may be approached cautiously as we try to get "a better feel" for them.

We should not regard this as a one-way street, however, and assume that feelings always follow thoughts. Our emotions play a vital role in shaping our judgments about people, places or objects. The "feeling" that something or someone elicits becomes part, often the crucial part, of the information we draw on in almost all the evaluations we make, and those who ignore their feelings in such matters do so usually to their own detriment.

Emphasizing the mental aspects of emotional experience should not necessarily mean downplaying the biochemical side of things. Yet this did happen for a while, because the Schacter-Singer results were accepted too uncritically. For one thing, adrenaline did not make the students any more happy than giving them saltwater. Thus adrenaline arousal might not really be a chemical enhancer or model of pleasure or happiness. The basic conclusion of the study — that there is no biochemical difference between various emotional states — is therefore questionable.

One reason for discussing Schacter and Singer's work in such detail is that it has been tested in regard to romantic feelings. One group of investigators studied young men who were walking across a high, rickety bridge. A female researcher approached them on the bridge and asked them to write stories based on a picture of a young woman covering her face with one hand and reaching with the other. Their sexual arousal was measured by story content and whether they accepted the researcher's invitation to call at a later time to discuss her project further. Their reactions were then compared with those of other men who encountered the same woman on a lower, more stable bridge. If changes in body chemistry for different types of emotions were the same, the researchers reasoned, then the students on the high bridge, who were assumed to be more anxious and hence thought to be putting out more adrenaline, would also feel more sexually aroused by the interviewer than would their presumably less anxious counterparts on the lower bridge, which in fact they were. As a further test, men still on the high bridge were compared to those who were interviewed ten minutes after they had crossed that bridge. Again the presumably most anxious group, i.e., those still on the bridge, wrote stories with more explicit sexual themes and had some members who called the interviewer after the experiment.

The main problems with these experiments are the "shaky" assumption about what people feel emotionally as they cross the bridge, and whether they view being invited to call later as a sexual come-on!

For those people on the high bridge who were truly scared, meeting the interviewer may have helped allay their anxiety. This in fact was demonstrated by the same researchers in another series of experiments that took place in a laboratory rather than on a bridge. It makes more sense to me that lowering anxiety might help stimulate sexual interest, although I think the pleasurably exciting aspects of the high bridge go furthest to explaining what was found.

What the researchers seemed not to have asked themselves is: Why are these people crossing that bridge? If you had taken people randomly off a city street and placed them on the high bridge, then you might see anxiety. Or even better, taken people who are scared of heights. I don't think you could even interview these people, let alone detect sexual arousal. But people who voluntarily seek out that bridge do so for the thrill and the excitement, which in some way must be pleasurable. If they feel a bit anxious as well, it is a secondary feeling spilling over from the uncertainty and sense of danger, which again they presumably find pleasurable. The men were asked if crossing the bridge felt scary, but not whether they were enjoying their crossing or would do it again. I remember being on a roller coaster once and feeling so terrified that I swore if I got off alive, I'd never go on one again. There is no way you could have detected any sexual arousal on that trip no matter who asked me a question, because it was sheer terror. But for people who were enjoying the ride, or the bridge crossing, it's certainly plausible that their capacities for sexual arousal would also increase. Thus the experiment is most likely showing not that anxiety heightens sexual arousal, but that other forms of excitement do so, which makes more sense both physiologically and experientially.

More recently, a number of sesearchers have tried to replicate the Schacter-Singer experiment and have been unable to do so.

My own impression is that the Schacter-Singer experiment was important because it highlighted the dual nature of emotional experiences. On the other hand, it is not at all clear that the physical aspects of all emotions are the same. In fact, strong evidence exists to the contrary. While our mental processes do play a large part in what we feel, our physical and biochemical processes are much more complex than was understood at the time of Schacter and Singer's experiment.

CHAPTER 2

EMOTIONS:
How We Feel What We Feel

If you were to leaf through some modern textbooks, you might get the impression we know a great deal about the workings of the human brain. You would find descriptions of brain architecture, theories about brain chemistry, printouts of electrical impulses, and detailed studies of the effects of brain injuries. However, while it is true we know much more about the workings of the brain than we did a generation or even a decade ago, it is also true that some of the most fundamental questions have not yet received anything like fundamental answers.

The workings of the brain presents us with a rather mysterious question — how do ethereal things like thoughts or feelings relate to electrical and chemical impulses? How does a three-pound mass of soft rubbery matter allow us to create symphonies, build bridges, or fall in love? By combining current knowledge with a liberal amount of informed speculation, I'm about to describe a model that, perhaps for the first time, links our thoughts and feelings to the workings of our brain in the area of romantic feelings.

THE NERVOUS SYSTEM

We need to start out with a brief, nontechnical description of how the nervous system works; you don't need to be a scientist to understand it.

Your nervous system has two parts to it — a central part, consisting of your brain and spinal cord, and a peripheral part — the rest of your nerves and sense organs. The whole apparatus has three main functions — to receive, transmit, and process information.

The building block of the nervous system is the nerve cell — a microscopic structure that consists of a cell body, a network of tentacles called dendrites that bring information in, and a long extension, called an axon, which carries information toward the next cell in the chain. Groups of axons banded together make up the nerves that run throughout the body.

Let me give you an example of how your nervous system works. You are in the kitchen cooking dinner and you grab the handle of a pot, momentarily forgetting that it has been over a flame for some time and is very hot. As you grab it you remember, and quickly let go, too late, however, to avoid getting burned. The actual pain comes a few seconds later, by which time you have already run to the sink to put your hand under a stream of cold water. A few minutes later, as the pain subsides and the first blister begins to form, you may say to yourself, "How could I be so dumb?"

Most of your nervous system is involved in this sequence. The decision to reach for the pot originates in the frontal or planning area of your brain. A message, generated by your brain's motor area, travels down your spinal cord and out along the motor nerves of your arm to whichever hand you generally use to pick up pots. Your eyes monitor the situation, transmitting information about the scene back to the visual area of your brain. As your hand is guided to the pot, your brain suddenly registers the larger gestalt — that the pot is on a stove, that the flame under the pot has been lit for quite some time, and that the pot handle is likely to be very hot. It's not clear exactly what throws the alarm — the visual area of your brain generating a mental picture of you with a hot pot in your hand, the sensation area of your brain creating the feeling of extreme heat in your hand before you actually touch the pot, or the general alarm area being triggered to create a momentary sense of panic. In any event the movement area of your brain gets the message, and an urgent new signal goes out: "GET AWAY FROM THAT POT." Depending on when you first register the danger, and how quickly your brain can send out a new signal, you may pull back before touching the pot or just after. Or, if your signals get a little jammed up, you may actually stand there for a moment with the hot pot in your hand.

If you've gotten burned, a new sequence begins. Heat-sensitive receptors in your hand become wildly activated, and transmit messages along the sensory nerves in your arm to your spinal cord and up to your brain that something very hot has been encountered. The message about temperature actually goes to the sensation area of your brain, while the message about pain registers somewhere else.

As you stand with your hand under the cold water faucet you have time to mentally replay the scene, again calling on the visual areas of your brain. Most likely, you also feel dumb and unhappy for having grabbed the pot, which means that your limbic system — the brain network responsible for emotional experience — is called into the act.

CHEMICAL MESSENGERS

From the standpoint of emotions, the most important thing about our nervous system is that nerve cells do not actually connect one with another. Rather, they are separated by minute spaces called synapses. Nervous system signals run the length of a cell, but must then be transmitted across the gap or synapse to the next cell in line. This task is handled chemically by our bodies.

When a nerve signal reaches the end of one cell, a small packet of one or another chemical substance is released from the end of that cell. The chemical molecules move across the synapse like a fleet of small boats and stimulate receivers on the edge of the next cell in line. This in turn stimulates the second cell to send a signal along its length.

There are approximately thirty known or suspected different chemical messengers, called neurotransmitters, that have currently been isolated, and new substances are being discovered all the time. From 1965 until recently, biological psychiatric research focused a great deal on neurotransmitters, because it was believed that fluctuations in their levels caused the profound emotional alterations that characterize many psychiatric conditions.

One of the principal neurotransmitters in the human brain is norepinephrine, which is closely related to adrenaline. This is an uncomplicated chemical that our bodies manufacture from a substance found in many foods and beverages.

For a long time we believed that depression resulted from a deficiency of norepinephrine needed for transmitting between nerve cells, while the opposite condition, mania, was due to having too much norepinephrine and hence too much nervous system activity. However, it now looks as if things are more complicated.

Another neurotransmitter, dopamine, which is quite similar to norepinephrine, seems to be involved in psychosis. One piece of evidence for this is that all of the drugs that are effective in treating psychosis block the activity of this neurotransmitter. This suggests that the presence of too much dopamine somehow causes hallucinations or delusions.

The problem with these theories is that no one has ever actually demonstrated an abnormal level of brain transmitters in depression, mania, or psychosis. More recently, attention has shifted to the role of the receivers (called receptors) that lie on the end surface of nerve cells and are stimulated by the chemical neurotransmitters.

CHEMICAL RECEIVERS

The receivers that lie on the far side of a synapse and are stimulated by neurotransmitters are called "postsynaptic receptors." If you are traveling from New York City to New Jersey, and think of the Hudson River as a synapse, then the New Jersey side would be the postsynaptic side. Actually a trip from Manhattan to Staten Island is a better example, because there still exists a ferry boat service between them, and this is a better model of how the brain transmits across synapses. One difference between rivers and synapses, however, is that you can cross a river in either direction, whereas chemical transmission across synapses only goes one way — from presynaptic to postsynaptic.

Receptors seem to receive only one or another type of chemical messenger. Thus there are specific dopamine receptors, norepinephrine receptors, and receptors for the numerous other neurotransmitters. A chemical that fits into a receptor can either stimulate it or block it, just as a key can open or lock a door. If enough receptors are stimulated, the nerve cell to which they are attached fires a signal which then travels down its length to the next synapse in line. On the other hand, if a blocking drug attaches to the receptors, the nerve cell becomes temporarily incapable of transmitting anything.

In addition to being stimulated or blocked, receptors can also become more or less sensitive to their particular chemical "keys." If a set of brain receptors become less sensitive, then more neurotransmitter is needed to activate them, and nerve system transmission slows down. On the other hand, if the receptors become more sensitive, that part of our nervous system "revs up." One theory that is currently being explored is that when someone becomes depressed and shows physical and mental sluggishness, what has happened is that his brain cell receptors have become less sensitive. This in turn would make them harder to stimulate chemically, thereby slowing down brain activity. In support of this, recent studies are suggesting that all available antidepressant drugs are capable of making certain receptors more sensitive, thus stimulating brain activity. This may be how antidepressants work.

THE DRUGS WITHIN US

In the last few years scientists have found that there exist specific receivers in the human brain for such drugs as heroin, opium, and Valium. This immediately led brain researchers to ask, "Why does the human brain have receptors for narcotics and tranquillizers?" After all, we have only been medicating ourselves for a few thousand years, so it's hard to imagine that the brain has evolved these receivers in that period of time.

If these receptors have not evolved just so we can get high on opium, feel pain relief from morphine, or relax with Valium, then what are they doing in our brains? Could it be, scientists began to ask, that there are natural substances in our bodies that can fit these chemical "locks" or receivers? Put another way, do our bodies contain naturally occurring narcotics or antianxiety agents?

In fact this has turned out to be just the case. In 1977 the first naturally occurring narcotic was identified. Since then more have been discovered. Depending on their size, they are called endorphins or enkephalins. While their functions are still not clear, some of them have painkilling properties that far exceed those of medically used narcotic drugs.

Naturally occurring chemicals that fit the same brain receptors that Valium fits are also beginning to be found. While their natural role is not clear, there are, at this time, two opposing schools of thought. One holds that the chemicals act like Valium, and help calm us down when we are under stress. However, at least some of the natural chemicals that fit these receptors seem to block rather than stimulate them, which could actually make us more rather than less anxious. Some scientists have wondered if this could be part of some built-in alarm system that helps arouse us in dangerous situations.

Amphetamines are powerful stimulating drugs that improve mood and increase energy for most people who take them. To date, no specific amphetamine receptors have been found in the human brain. Instead, it seems that amphetamine affects our nervous system indirectly — by increasing levels of norepinephrine and dopamine. However, scientists have also identified in the human brain a naturally occurring amphetaminelike substance called phenylethylamine (PEA).

Some very interesting findings have come out lately. Drugs that raise norepinephrine, dopamine, and PEA levels at times cause overstimulation, in which people need only a few hours' sleep, tend to feel very optimistic about the future, and may even become more socially outgoing than is usual for them. Interestingly, this is similar to what happens to people when they get promoted, win a lottery, or fall in love. This suggests that happy events or pleasant thoughts exert their mood-lifting and energizing effects on us by increasing the activity in our brains of PEA or some other naturally occurring amphetaminelike substance.

The flip side of this hypothesis is that the lowered mood and decreased energy that come from unpleasant memories or experiences, such as romantic disappointment, are due to lowered levels of some brain chemicals or diminished sensitivity of key receptors. Again this could be due to diminished internal brain amphetamine activity. Please keep in mind that so far these are hypotheses and speculations, not proven facts. However, data beginning to emerge from a variety of research projects suggest that there are amphetaminelike fluctuations in our brains in response to such things as romance, applause, and rejection. Before elaborating how this might happen, we need to look a bit further into how our brains work.

OUR EMOTIONAL WIRING

While stimulant drugs like amphetamine may affect many brain areas, the emotional changes they induce are most likely due to their effects on the brain's limbic system, that area of the nervous system primarily responsible for our emotional experiences. In the early part of this century there was much speculation and disagreement among scientists as to what part of the brain was involved in emotional experience.

One way to visualize the human brain (see the accompanying illustration) sitting in our skull is to crack a walnut and remove

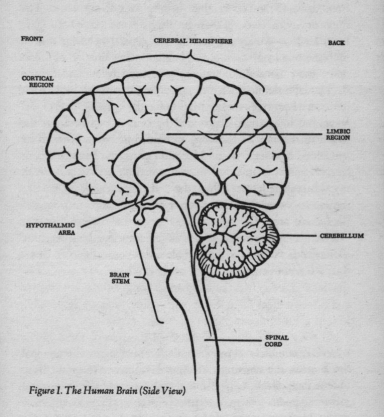

FRONT   CEREBRAL HEMISPHERE   BACK

CORTICAL
REGION

LIMBIC
REGION

HYPOTHALMIC
AREA

CEREBELLUM

BRAIN
STEM

SPINAL
CORD

Figure I. The Human Brain (Side View)

the top half of the shell while leaving the nut inside undisturbed. If you look down at the two hemispheres of the nut with the fissure dividing them you have a good approximation of the view from above of the two cerebral hemispheres of our brain. The upper or cortical layers of these hemispheres are the site of higher brain activity — planning and certain general behavior control up front, fine motor control just behind, discrimination of sensation toward the middle, auditory processing below and behind that, visual discrimination and processing toward the back. Each hemisphere provides these activities for the opposite half of the body, and integration is provided by connections between the two brain halves. In addition, each hemisphere appears to be specialized in certain ways — for most of us, the left side of our cerebral cortex is involved in mathematical, verbal and other, more logical activities, while the right side excels at pattern recognition, spatial relations, and perhaps more artistic and mechanical pursuits.

The upper or cortical area is highly developed in primates, especially so in humans. Lying below this cortical mantle, however, is a set of structures that is rather similar in all mammals — the limbic system. It is the interconnections of the limbic system and adjacent areas (like the hypothalamus) that are responsible for our abilities to experience emotions.

An early set of studies highlighted the role of these limbic structures in emotional response. Different changes in emotional behavior were demonstrated in animals subject to surgical cuts at various brain levels. Cuts below the uppermost brain layers, but above the limbic system, disconnect the cortical region but leave the limbic system intact and still connected to the rest of the nervous system. Such surgical cuts produce animals who can be easily provoked to extreme emotional responses, especially rage. Typically, however, these animals could not direct their rage toward any particular objects. On the basis of experiments such as these it seems that the uppermost brain layers serve to inhibit emotional centers deeper in the brain, and also to direct emotional responses. On the other hand, cuts made more deeply into the limbic region tend to diminish the emotional response itself.

One other set of animal experiments must be mentioned because of their potential significance for understanding human emotional experience. In the early 1950s, two brain researchers named James Olds and Peter Milner were investigating whether rats might be made uncomfortable by electrical stimulation of a certain area of their brans. Accidentially the tip of the electrode was placed slightly off target. An electrical current was given to the rat whenever it entered a specific corner of a box, with the theory that if the effect was uncomfortable for the animal, it would begin to stay away from that corner. Instead, it came back quickly after the first stimulation and even more quickly after the second, suggesting that the rat found the electrical current pleasant and wanted more.

Olds and Milner then implanted electrodes in this area of the brain in a number of rats and allowed them to stimulate themselves by pressing a lever. The animals pressed the lever over and over again, as much as seven hundred times per hour in the first experiment, and up to thousands of presses per hour in later studies. Quite obviously, electrical stimulation to specific areas of a rat's brain is highly rewarding and generates a desire for further stimulation.

These areas of the rat's brain that crave self-stimulation have been called "pleasure centers." Located in the limbic system they are thought to be part of the brain's normal reward system. Other limbic brain areas are thought to contain "displeasure centers"; wired animals will try to avoid anything that results in stimulation of these brain areas.

Interestingly, amphetamine, which most human beings find pleasurable, has been shown to facilitate self-stimulation of the "pleasure centers" in rats. That is, rats wired to receive low current stimulation of their pleasure centers will press the self-stimulation lever more frequently if first given amphetamine than if not given the drug. While discrete "pleasure centers" have not, up to now, been conclusively identified in humans, a logical question to ask is: Does the "pleasure center" concept offer us any clues to understanding human emotional behavior?

The best way to approach this is to ask: What makes people do any of the things they do?

At first glance this may seem a silly question, because human beings appear to be motivated by many factors. For reward, some individuals are motivated by desires for wealth, fame, recognition, human service, salvation, love, sex, adventure, escape from routine, and so forth. On the negative side, they seek to avoid things like poverty, jail, obscurity, physical discomfort, ostracism, boring routine, eternal damnation, and so on.

Taking this discussion to another level, however, we can ask: What makes the items of the first class rewarding or those of the second class aversive to most people?

The answer is that objects or experiences are sought or avoided because of how they make us feel. Or more precisely, how the idea of having them, in the case of objects, or being in them, in the case of experiences, makes us feel.

These feeling states that are provoked by our thoughts, memories, or ongoing experiences are crucial, or perhaps *the* crucial determinant of human behavior. If the idea of pursuing or attaining a goal makes us feel excited, then chances are we will be highly motivated to work for this goal. Things that elicit less excitement will be generally less motivating. Decreasing our guilt or anxiety, avoiding shame or humiliation, and reducing worry or fear are also powerful motivators of human behavior, as most of us realize when it's time to pay our taxes or hand in an assignment.

At this point some people may want to turn to sociology or psychology to explain motivation, but having gone through our discussion of brain function and chemistry, we can ask the next logical, and to me the really central question: How do thoughts, fantasies, memories or experiences elicit pleasant or unpleasant emotions in us? What happens in our brains so one experience or memory elicits warm, glowing feelings and another painful embarrassment? Why do some things call forth very powerful feelings, while others register only mild reactions, and still others no emotional response at all? And why, on some days, do many thoughts or encounters feel exciting, while on other days the same events do not arouse comparable emotions?

The answers seem to lie in our brain chemistry. For a thought, memory, fantasy or experience to evoke a feeling of pleasurable excitement it must stimulate a pleasure center in the brain's limbic system (keeping in mind that we don't know if human pleasure centers are really specific brain areas or part of a widespread circuit). Similarly, anything that elicits negative feelings is, in some way, stimulating the brain area associated with unpleasant feelings (anxiety, sadness, etc.), or is turning down the activity or accessability of the pleasure centers, or both.

Thoughts, experiences, or memories that do not activate your limbic system are experienced as neutral, unemotional or simply boring. On the other hand someone can feel "overstimulated" by too many good things happening too quickly or by too many exciting ideas, in which case it appears that excitation sometimes spills over from the pleasure center to stimulate anxiety or some other unpleasant feeling. In most but not all circumstances the two centers appear to act reciprocally: if the pleasure center is firing, it is harder for displeasure circuits to be simultaneously triggered; while if unpleasant emotions are being elicited, the pleasure center seems to become harder to stimulate at that time.

This model can explain the differences between sad and happy thoughts, experiences, and memories, between optimistic and pessimistic people, between mania and depression, and between the good and bad days or times in our lives. How a thought connects to your pleasure or displeasure system will determine what emotion it arouses. Memories that are always pleasant to recall — whether of a wonderful vacation, a romance, a beloved relative or cherished pet — can be thought of as having well-established connections to your pleasure centers. Anything that evokes the memory immediately triggers the good feelings.

Similarly, painful memories have "well trodden" pathways to brain areas associated with displeasure, anxiety, and so on, such that anything that evokes these memories brings on painful feelings.

It is remarkable how long things like this can endure. Many adults have had the experience of a distant memory from childhood suddenly evoking a wave of guilt or intense shame. Overall, the link between emotions and memories is a crucial one from an evolutionary and daily survival standpoint, since it permits us to recall how we felt the first time around, and to pursue experiences that were rewarding and avoid those that were painful or frightening.

EMOTIONAL PROGRAMMING

At this point I hope it is starting to become apparent what all this has to do with love. If not, it will become very clear shortly.

Each of us carries a store of recollections about our spouses, relatives, friends, children, parents, grandparents, other care-takers, teachers, and childhood playmates. Our senses of identity and self-worth are thought to be shaped by our views of ourselves in all these relationships, as are our attitudes toward and interest and trust in other people. Memories of these important relationships are stored in the brain's data banks, which is how specific relationships can live on in our minds, sometimes out of awareness, until triggered by an encounter that reactivates an old memory. When we experience very strong reactions to people we barely know, it may have to do with that person's rekindling an old memory, which elicits the strong feeling that pertained to that earlier relationship, and may have nothing to do, really, with this new person.

Psychoanalysts have generally been interested in our memories of important relationships as a way of understanding our views about ourselves and others. What has received less attention, but may be as important, is the link between memory and our discussion of the brain's emotional centers.

Simply put, a positive memory of yourself in relationship to someone else is a thought that has established a firm link to your pleasure center. Similarly, a memory that calls up painful emotions is linked to a displeasure circuit. The importance of this is that we can now understand in biological terms how good and bad early life experiences affect our personalities, as well as ongoing experiences of life.

If someone grows up in a cruel, neglectful, uncaring or cold family atmosphere, the chances are that as an adult he or she is going to have a store of painful memories. What this means is that there will be a series of well-established links between memory and displeasure centers. As adults these people will be more prone to depression, sadness, or pessimism. Anytime an unhappy childhood memory is evoked, the displeasure circuits will be activated. Also, new interactions that touch on any of these old memories will often cause excessive emotional reactions in people who may not even be consciously aware of the connections between present and past. This is what leads psychotherapists to probe childhood experiences whenever they see an adult emotionally overreacting to certain kinds of situations.

Keep in mind, however, that not all people are affected the same way by similar childhood experiences. While this has often been attributed to subtle differences in their actual experiences, it is possible that differences in brain biology account for similar experiences having dissimilar impacts in various individuals. We know that newborn babies have different temperaments, and that certain emotional tendencies can be inherited. It may be that we are all born with different pleasure or displeasure thresholds. A temperamentally lively or active child could have pleasure circuits that are more easily or strongly stimulated and harder to shut down, or displeasure circuits that take more to make them fire or fire less intensely. For others, painful experiences may register more easily.

The sum total of our biological temperament and experiences causes all of us to be programmed or "wired" in our own individual ways. While much of how we feel in day-to-day life will be determined by our ongoing experiences, how we are individually wired will also play a role in several crucial respects. The more optimistic or pessimistic types among us may be at opposite ends of the spectrum in how easily their pleasure or displeasure centers are activated or shut down. In this sense an optimistic person is someone whose pleasure centers are activated more readily by new thoughts or plans for the future, while a pessimistic person has more sluggish pleasure centers and/or more easily activated displeasure circuitry. Second, those with storehouses of pleasant early memories will constantly be remembering things that trigger pleasure responses, giving them repeated lifts of mood and energy. The same phenomenon, with opposite, more depressing effects, occurs for people with the most unhappy childhoods. Third, and most important, an individual's pleasure and displeasure centers do not appear to have fixed thresholds; rather, they seem continually to vary over a range that seems to be different for each person.

What makes your limbic pleasure center threshold vary is not fully clear, but experience tells us that a number of things affect it. For one, it often seems to act as a positive feedback system — if you are having a pleasant memory or experience, you are more likely to experience the next thought or encounter in a similar way. Highly pleasant experiences, such as a job promotion or falling in love, can reset the whole system at a different level, so for a while everything may feel more interesting and exciting.

On a good day, whether it comes out of the blue or because something pleasant happened, your pleasure center threshold appears to be lower, and more thoughts and experiences cause the center to fire. One crucial aspect of a happy childhood and a rich store of positive memories is that you are constantly being activated by pleasant recollections of the past.

All this can be said as well about negative experiences. Lose a job or break up with your lover and everything turns to gloom. It is harder for a new project or new person to make you feel very good, at least for a while. In such a situation your pleasure threshold has been turned way up, so little or nothing can get through. People in whom painful childhood experiences are constantly stirred up are experiencing bombardment of their displeasure centers, rendering them more vulnerable and more likely to experience new events as painful as well.

In general, experiences seem to have greater impact on us than do thoughts, fantasies, or memories. Many states of depression are characterized by feeling unhappy, unstimulated, and pessimistic when one thinks about one's life and future possibilities. At such times, thoughts, memories, and fantasies are clearly not triggering the pleasure centers. A well-known treatment for feeling "down in the dumps" is to go out and have some fun, i.e., to put oneself in a potentially pleasurable situation, be it a party, a funny movie, or a relaxing vacation. What one is doing in such a situation is "priming" one's pleasure center by getting it to fire again, and hopefully resetting its threshold at a lower level, so once again thoughts, plans, and other mental activity will feel exciting.

Unfortunately there is one form of depression that is not helped by pushing oneself to do things one normally enjoys. In such a state, which we call melancholia, people become incapable of experiencing any pleasure or joy — there is a total shutdown of brain pleasure centers. This can have grave consequences, including suicide, which may feel quite reasonable under the circumstances, although it is really not at all so, because effective treatments exist that can quickly restore a melancholic individual's normal capacity for pleasure and joy.

It should be clear by now just what all this has been leading up to. Love and romance seems to be one, if not the most powerful activator of our pleasure centers. Falling in love involves creating a new relationship. It exists in actual fact, and also for you to carry around in your mind. Both tend to be very exciting emotionally. Being with the person or even just thinking of him or her is highly stimulating. If the relationship is not established or is uncertain, anxiety or other displeasure centers may be quite active as well, producing a situation of great emotional turmoil as the lover swings between hope and torment. But if the relationship is going well, everything in life for a time takes on a rosy hue, which can be seen as a lowered pleasure threshold for other mental processes or daily experience.

One is tempted to ask: Why should love have this effect on us, or on our brains? Love is, by definition, the strongest positive feeling we can have, the strongest arousal of our limbic pleasure center that can be induced by relating to, being with, thinking of or remembering another person. Other things — stimulant drugs, passionate causes, manic states — can induce powerful changes in our brains, but none so reliably, so enduringly, or so delightfully as that "right" other person.

The bottom line is that our brains are wired so as to "light up" very strongly when certain things happen between us and another person. This is something that has evolved in us as a species over many thousands of years, and seems to be nature's way of ensuring that we seek out and form relationships that last at least long enough to reproduce a next generation.

But such a view raises many questions. Could our cravings for romance, for example, really be our seeking a particular kind of high? Could those who fall in love repeatedly, especially with the "wrong" people, have pleasure centers that are wired to light up too readily? Could the fact that most romances lose some of their intensity with time be due to some biological process that occurs within us? The implications of seeing love, basically, as a "glow" in our pleasure centers brought about by another person are enormous.

CHAPTER 3

FEELING GOOD:
How Our Pleasure Centers Work

Mrs. T. called me at 11 P.M. on a Thursday. Her husband, a successful banker in his early fifties, was talking about suicide. Would I please see him as soon as possible?

When Mrs. T. brought her husband in to see me the next day, he certainly didn't look like a successful banker. He looked haggard, disheveled and unshaven, and his face showed little expression as he talked. In the ninety minutes I spent with him he never smiled.

Mrs. T. went to some pains to tell me that her husband wasn't always like this. Usually a confident, gregarious and witty man, he had started becoming depressed several months earlier, possibly because his bank was losing money. He began to eat less, started waking up too early in the morning, and seemed gradually to lose his zest for life. In the last week he'd stopped working, and had to be pushed to bathe, shave, or change his clothes.

Mr. T. seemed to feel that his world had come to an end. He was convinced his bank would fail (it wasn't even close to failing), that he was responsible, and that there was absolutely no reason to go on with life. Everything seemed absolutely gray and gloomy, and nothing gave him the slightest lift.

I've gone into some detail about Mr. T. because I want to give you a feeling of what happens when a person's pleasure centers shut down. Mr. T. had slipped into a severe depression. What this means, in essence, is that his brain chemistry had become unbalanced. Thus nothing — no thoughts, no activities, nobody — could give him the slightest feeling of joy or hope, because nothing could make his brain pleasure centers light up.

What I did for Mr. T. was to make sure that he wouldn't be alone for the next several weeks (because he was a real suicide risk), then put him on antidepressants. Two weeks later he was starting to improve. In six weeks he was back to his old self.

Despite the fact that modern psychiatry knows how to treat severe depression, we actually know very little about the brain's pleasure centers. In part this is because they're hard to study. But also I think that their absolutely central role in our emotional lives has not been fully understood.

As a bridge between the previous sections on emotions and brain chemistry and the chapters on romance that follow, this chapter looks at new ways to tell how our pleasure centers are working, how they work, and how people differ biologically in their capacities for excitement and joy.

MEASURING OUR PLEASURE CENTERS

If we want to know more about how our pleasure centers work in response to romance or other kinds of experience we need ways actually to measure how these brain areas are functioning. What follows are descriptions of three ways that scientists are trying to do this.

One very controversial project has studied the electrical behavior of the human brain during various emotional states. What has made this work unique is that, while most electroencephalographic studies (EEG) measuring brain waves are done by electrodes placed on the scalp and measure only the electrical activity reaching the brain's surface, this project has studied human brain activity via electrodes placed deep in the brain's limbic system. Deep brain center activity cannot be recorded by scalp electrodes. The reason why limbic system electrodes are not used routinely is that, to place them in position, neurosurgery is required.

As I will discuss shortly, some new developments may soon make such an approach unnecessary. But what these investigators have reported is that a certain area of the brain (called the septal region) seems to show a characteristic pattern of electrical activity during a variety of euphoric states, including those induced by smoking marijuana and by sexual orgasm. Moreover, they claim that pleasure responses have been elicited by electrical and chemical stimulation to this region, and that patients with electrodes there will stimulate themselves three hundred to five hundred times an hour if given an opportunity. Also, impaired electrical activity of this septal region has been associated with low mood states. Moreover, destructive cuts to this brain area in animals have impaired emotional expression.

Since it is difficult for any investigators to try to monitor our deep brain structure directly, we have to look for some more accessible way to measure the brain's pleasure circuitry. Surprisingly, one that has been studied in a very limited way confronts us every day when we look in the mirror. What I am referring to is the human face.

Our faces, and particularly our mouths, quite accurately reflect a wide variety of our emotional states. What we react to and how much facial expression we show vary across individuals and especially across cultures. Nevertheless, the basic facial patterns — the smile, the frown, the pout, the snarl, the look of shock or surprise — appear to be universal human traits. Moreover, many are involuntary — they occur without thinking and are not controlled by the cerebral cortex. Someone can have a stroke that impairs his ability to voluntarily move his mouth into the shape of a smile, yet he can involuntarily smile in response to a joke. It has been suggested that the mouth in particular and face in general are "hard wired" (directly and permanently connected) to the brain centers that regulate emotion.

One could simply photograph people's facial expressions in response to different activities and develop some measuring system. But a more accurate method exists — namely electromyography, or EMG. This is a recording device that measures the electrical activity generated by muscle contractions, using electrodes pasted on the skin or tiny needles painlessly inserted into particular muscles. Some of the muscles that contract when we smile are different from those that make us frown, so electrodes placed in or over the relevant muscles could measure the amount of smiling or frowning activity in a particular time period. Someone's ability to experience pleasure, or for that matter any other emotional reaction, in response to a variety of stimuli such as pictures, films, drugs, or whatever could then be measured by recording the pattern of facial muscle activity as that person engages in a particular activity.

As far as I can tell very few researchers have tried to use this technique to look at human pleasure center activity. One possible application might be to measure how much people smile in response to humor or comedy of some kind. While people obviously differ in what they find humorous, this would be a reliable way to track changes in a particular person over time. This could be especially important in someone who periodically becomes depressed, since we could assess the extent of pleasure center shutdown during a depressed period and also measure when that person's pleasure capacity was fully recovered.

A third way to look at our pleasure centers is with a new science-fiction-type device, called the positive emission tomography or PET Scanner, that is now being used as a research tool in several U.S. medical centers. A generation ahead of the recently introduced CAT Scanners that have revolutionized our X-ray procedures, the PET Scanner involves giving a person a single dose of a radioactively labeled substance that will be absorbed by the brain. The substance that is used is deoxyglucose, a close analogue of glucose, or sugar. Glucose is taken up by the brain and used fully for energy, and its by-products are rapidly transported out of the brain again. Deoxyglucose, on the other hand, enters brain metabolic pathways and quickly gets stuck. This happens because brain enzymes are not equipped to deal fully with this unnatural form of glucose. If a short-lived radioactive substance is attached to the deoxyglucose, this radioactive "label" will also get stuck in that part of the brain that has taken up the deoxyglucose. A photographic scanner, sensitive to radioactive emissions, can then film the brain and locate all the areas where the radioactive, or "hot," deoxyglucose has been taken up. We can then construct a picture of the brain's activity pattern, since the brain areas most in use at the moment will have taken up the most deoxyglucose, just as they normally take up the most glucose to fuel their high level of activity.

What this all means is that the areas of the brain that participate in any given activity, be it an anxiety attack, a seizure, or an orgasm, can be mapped out with the PET Scanner. Using radioactive chemical·tags may sound dangerous, but those actually employed have short half-lives, so that they decay rapidly and then cease to be radioactive. The problem this creates is that each PET Scan center must have its own cyclotron, a huge facility that can generate the radioactive substances, because after they are produced they must be rushed over to the X-ray department, quickly injected into the subject, and then "photographed," all within thirty minutes or so before the radioactivity wanes. In practical terms, a PET Scanner and its accompanying cyclotron cost something in the range of several million dollars, which is why only a few centers have them. In addition, the kind of pictures you now get for this money are somewhat reminiscent of those produced by a 1948 television set. That is to say, the technology still has a way to go. But future possibilities boggle the mind. It's not unreasonable to think that at some not-too-distant date one will be able to get a PET Scan to help with the diagnosis of various mental disorders that may be related to abnormal limbic or other brain area activity patterns. We could also use the machine to see how our pleasure centers or other brain areas were functioning during any particular kind of activity or emotional state.

But while the PET Scanner, or for that matter deep brain electrodes or measuring smiles, may tell us something of how our pleasure centers are working, they don't tell us much about how they work. On the other hand, mind-altering drugs tell us a great deal about how our pleasure centers work.

WHAT DRUGS TELL US ABOUT OUR BRAINS

With the upsurge in nonmedical use of mind-altering drugs over the past twenty years, feeling good or "high" has acquired a bad connotation. What I hope to convey is that from the standpoint of brain biochemistry, as opposed to philosophy, law, or even health, the distinction between drug- and nondrug-induced highs has been overdrawn. While the things that cause human beings to feel intense pleasure, excitement, or euphoria are incredibly diverse, they all ultimately operate by inducing certain changes in our brains. This holds true whether the stimulus is making a new friend, climbing a difficult mountain, building a financial empire, creating a work of art, or snorting cocaine. There has been a tendency to see drug highs as exclusively chemical and, therefore, different from nondrug highs, without understanding that nondrug experiences also operate by inducing chemical changes in us.

A good example of how a nonchemical experience can induce chemical changes in the body is the effect of placebos on pain. A placebo, which is a chemically inert substance, can often alleviate severe pain for several hours at a time. This has always been thought of as a psychological effect, and often hospital patients who benefited from placebos were thought not to be having real pain.

However, a recent very important study throws a whole new light on placebo effects and suggests that these inert substances exert their effect by inducing chemical changes in our bodies.

In Chapter 2 we spoke of the endorphins and enkephalins, narcoticlike substances manufactured by our bodies. Could it be, scientists began to wonder, that the painkilling effects of placebos are due to their ability to affect endorphin levels or activity in our bodies? One way to test this hypothesis is with naloxone, a narcotic antagonist that blocks the effects of both ingested narcotics and also our internal narcotics. Volunteer patients having molars removed in a dental clinic were given inactive placebos (after being told they might receive a narcotic, a placebo, or a narcotic blocker). More than a third of those who received a placebo experienced great pain relief. An hour later, half of these people got a shot of naloxone; the other half a second placebo. The naloxone brought the pain back, while those who received a second placebo instead of the naloxone continued to have pain relief. What this experiment suggests is that placebos help some people to deal with pain by stimulating their bodies to produce their own internal painkiller, an effect which was undone by the naloxone.*

We have some good evidence that our minds can be trained to raise or lower endorphin levels. For example, the taking of an inactive pill or use of a needle and syringe that contains no drug can produce relief of the withdrawal symptoms that narcotics addicts experience when they run out of drugs, presumably by stimulating the brain's own narcotic system to produce more (although this has never been tested with naloxone). Conversely, former addicts who are detoxified and uncomfortable without drugs in drug-free environments begin to experience craving and withdrawal symptoms when they return to an environment where drugs are available. These observations suggest that nondrug experiences actually have biochemical effects on us.

Numerous parallels exist between our emotional reactions to drugs and nondrug experiences. To me this suggests that many sorts of nondrug experiences induce chemical changes in our brains quite similar to what happens when we take certain drugs, and that these resulting chemical changes give rise to the specific emotional and physical effects we experience at any given moment. While we are far from a complete understanding of how the brain reacts to either drug or nondrug stimulation of different types, a number of things are known. The bottom line is that extreme biological change, whether drug induced, experientially caused or even "out of the blue," can lead to patterns or states of exhilaration on one hand and profound misery and intense emotional pain on the other.

The technical term for drugs that alter mental functioning is psychoactive. Psychoactive drugs can be said to exert either direct

---

* These findings must be taken as only tentative because several other research groups have not been able to replicate them.

or indirect chemical effects. Direct-acting drugs fit into brain receptors where they either mimic or block the effects of naturally occurring brain chemicals. As we discussed in Chapter 2, narcotics like heroin, opium, and morphine fit into opiate receptors, and produce the same effect as the brain's "own narcotics," the endorphins and enkephalins. Caffeine, on the other hand, may insert itself on or near the newly discovered Valium receptor but block rather than mimic the effect of our naturally occurring tranquillizers, which have yet to be completely identified. This blocking effect may explain caffeine's stimulating, and, at times, anxiety-producing effects.

Other drugs act indirectly by stimulating the brain to produce or secrete more of one or another neurotransmitter. If you want to see the effects of increasing the brain's levels of norepinephrine and dopamine, just take a look at someone who has taken cocaine or amphetamine. Antidepressant drugs may essentially do the same, but seem to work only in people whose brain chemistry is out of kilter to begin with. LSD, a highly potent psychedelic, affects a number of brain chemical systems, including serotinin, another wide-acting neurotransmitter. Some nonchemical physical stimuli act via this indirect mechanism. For example, both physical injury and acupuncture are thought to liberate brain endorphins and enkephalins, which may diminish feelings of pain. As I discussed earlier, placebos appear to do this for some people as well.

THE POWER OF SUDDEN CHANGE

A central rule of psychoactive drug action is that the human brain responds more to changes in level than to the actual level itself. What this means is that a drug has greater effects when its presence in the brain or bloodstream is rapidly increasing than when it reaches its peak amount and levels off. Similarly, we again see greater mental effects when the drug level begins to fall, although in this case the effect is usually the opposite of what was seen when the level was rising.

While a detailed description of how drugs are absorbed or metabolized in our systems belongs in pharmacology textbooks, a simple overview is in order. When a psychoactive drug is taken into the system, it enters the bloodstream, then crosses the membranes that separate our bloodstream from our brains (the "blood brain barrier") and diffuses into the brain. The faster the rise in the bloodstream, the greater the wallop suddenly delivered to the brain. Thus drugs injected into a vein have much more potent effects on us than similar-sized doses taken by mouth, because our brains are hit by a sizable amount of drug all at once. On the other hand, the effects of an oral dose last longer, because they take longer to build up.

Similarly, how fast a drug is withdrawn from the body in large measure determines how severely we will react. If someone who has been taking a steady amount of sedative or alcohol stops suddenly, he or she will experience disturbing "withdrawal effects," including restlessness and agitation, that are the opposite of the tranquilizing effects of the drug. On the other hand, if the dosage is gradually lowered over an extended period, the effect is much more gentle. Of course the amount being taken to start with and the length of time it has been taken are also important. Suddenly stopping a large sedative or alcohol dose to which someone has become accustomed can cause a severe reaction, including convulsions, while stopping a smaller dose or one used only briefly may only produce mild agitation.

Throwing someone a surprise birthday party may be like injecting him in the vein with a large-dose stimulant. He walks in the room, the lights go on, and *bang* — his brain is hit with a massive stimulus all at once. Tell him about the party two days before, and he will still be excited, but it's more like an oral dose of the same pill; the excitement builds up more gradually and may last longer, but it does not reach the same intensity.

Sudden losses in life may be experienced by our brains as a form of acute withdrawal. The unexpected loss of a job can be a brutal experience, and the surprise loss of a loved one is usually devastating. On the other hand, people who have been gradually warned, whether through hints from a boss about their job, or a doctor's report on the condition of someone close who is ill, usually do not experience the same intensity of pain and anguish, although again, it may be more prolonged.

GETTING USED TO SOMETHING WITH TIME

While we usually say that such people have had time to adjust, we usually mean this in exclusively psychological terms. But pharmacologists know that our brains adjust biochemically over time, which is why the gradual buildup or decrease of a drug affects us less intensely than do sudden changes. Constant blood and brain levels of many psychoactive drugs have less and less effect on us over time. This is due to the brain's adapting itself as it becomes used to having the drug around, and in pharmacological terms is called tolerance, meaning the brain learns to tolerate the drug. A possible mechanism for this involves changes in receptors. Chronic exposure to certain psychoactive drugs, or to an excess of the neurotransmitters they stimulate, seems to lead to a reduction in the number of receptors, and may be the basis for certain forms of drug (or nondrug) experiences losing their impact.

This explains why drug abusers either gradually have to increase their doses to get the same effect, or have to take holidays from their drug to let their brains become "unused" to it again, so that small doses will once more get them high. Tolerance or adaptation also occurs to a single dose of a psychoactive drug. If you inhale or "snort" a sizable dose of cocaine, the blood level begins to build rapidly and reaches its peak in about thiry minutes. This parallels both physical (increased heart rate) and emotional (feeling euphoric) effects of the drug, and users report their maximum "highs" fifteen to twenty minutes after inhalation. However, blood levels of cocaine actually remain quite elevated at least two hours after snorting it, while physical and emotional effects are completely gone by the end of sixty to ninety minutes.

Tolerance to nondrug stimuli may be an important but largely unrecognized aspect of human experience. Your first promotion, your first publication, your first child — these can be overwhelmingly powerful experiences. The second is also great, but a little less so, the third, yet less, and so on. What we try to do the second time to give us the same thrill is to increase the dose or vary the nature of the experience by asking for a bigger raise, trying for a more widely read publication, or hoping for a baby girl if we already have a boy.

Tolerance appears to develop in many unchanging situations to which we are constantly exposed. To some people this creates a feeling of security, but to others it is devastatingly boring, and creates a yearning for change or novelty. Novelty means something new, something which is not familiar, something to which one's brain has not become tolerant. The effects of tolerance on our personal relationships are discussed in the chapters that follow, but I don't think it is an accident that the great romances of literature, such as Romeo and Juliet, or Cathy and Heathcliff, occur between people who, for one reason or another, don't have a chance to stay together. Thus their brain pleasure centers have not had time to get used to, and therefore less excited by, having the other person around. As we know from experience, falling in love is very different from, and I think much easier than, staying in love. What has not been previously considered is that a large part of the difference has to do with the way our brain chemistry works.

Other examples abound, such as the need that many people have for the stimulation and variety provided by vacations, varied work activities, professional challenges, hobbies, etc. Similarly, it could explain why retirement, which many envision as a constant vacation full of golf or fishing, may become very dull, when the same pursuits were very satisfying for a few weeks at a time during a busy work life.

Interestingly, for reasons we don't fully understand, some people show "reverse tolerance" to certain drugs, which means that the same dose over time begins to have bigger rather than smaller effects. Certain cocaine and amphetamine effects show this reverse tolerance, and these effects may become exaggerated with chronic use. This mechanism may also operate in certain human relationships.

PHYSICAL DEPENDENCE, REBOUND, AND ADDICTION

Other important drug principles include physical dependence, rebound, and addiction. The degree of one's dependence on a drug is judged by the severity of the withdrawal symptoms that occur when that drug is suddenly stopped. As mentioned, individuals heavily dependent on alcohol can have convulsion or suffer delirium tremens (the DTs) if they suddenly stop drinking. Similarly, narcotics users go through intense, although not life-threatening, withdrawal, while stimulant abusers crash into depression and lethargy.

We don't speak of people becoming physically dependent on marijuana. However, some people do get dependent on marijuana in the sense that they use it regularly and compulsively, and go to great lengths to procure the drug if they run out. What seems to be going on here is that they are medicating themselves to cover up underlying states of anxiety, depression, or dissatisfaction, which begin to emerge whenever they run out of the drug. Physical dependence should not be equaled with compulsive drug use. Someone can be physically dependent on a drug, lose access to it, and go through withdrawal. After that he is no longer physically dependent. If at some future point he begins to seek or use the drug again, then he is manifesting compulsive drug use.

Human beings certainly become dependent on relationships, jobs, and so forth, as evidenced by the "withdrawal" they suffer if these things suddenly disappear. Whether to think of this as psychological or physical dependence, however, is unclear. If someone loses a job, there is a tendency to become depressed and discouraged for a time. Does that mean that the loss of the job induces biochemical changes in us, and that having the job maintained a certain chemical stability that was altered by being fired? In a sense this is true, but there are psychological variables such as what the job represented to the person (e.g., success, status), and whether losing it is experienced as an indication of incompetence, lack of ability, and so on. While this gets complex, there is no way to avoid it, because human beings cannot be reduced to simple biological reflexes and reactions. However, they should also not be treated as total ethereal, spiritual, or psychological creatures that do not have bodies and brains. In fact many people who lose a job or a loved one go into depressed states for which the treatment is often partly or fully chemical, which makes sense if we keep in mind that all emotionally intense experiences have neurochemical as well as psychological aspects.

Certain parallels with drug withdrawal appear to occur during temporary separation from a husband, wife, boyfriend or girlfriend with whom one is very intimate. Granted there are important psychological reactions, such as fear or feelings of abandonment. Nevertheless, what often happens is that one begins to miss deeply and, in fact, crave the other's presence. Feelings of anxiety may set in, coupled with a certain loss of energy, optimism, and enthusiasm, which has been likened to "running down one's battery." The thought of seeing the person again, a phone call or letter, and best of all, a reunion, can be exciting, calming, or energizing events, perhaps in part because of the physical changes they induce in us. As an example, while campaigning for the presidency, Ronald Reagan was described by his associates as looking as if he had received a "tonic" when his wife, to whom he is very attached, rejoined him after a short separation. What was being described was the added bounce, sparkle, and energy that he displayed when his wife returned and which had begun to wane in her absence. Whether such behavior is, in fact, biologically the same as drug effects, or only an unrelated analogy, will be central to our discussion.

Parallels also exist between other aspects of nondrug and drug behavior. For example, we often speak of someone getting married on the rebound, by which we mean that he or she made a hasty or inappropriate partner choice while still suffering the emotional effects of a recently ended romance. It so happens that periods of intense or excessive drug-induced stimulation are often followed by periods of emotional and physical depression, and vice versa. This pharmacological phenomenon is known as a rebound effect. In a sense the brain reacts like a spring that recoils forcefully if stretched too far or expands forcefully if compressed too tightly. The applicability of this biological model to romantic partner choice will be discussed in the chapters that follow.

Why people become drug addicts, which drug they abuse, and which people get hooked are complex issues with both psychological and biological components. In general, complusive drug users tend to be more impulsive, rebellious toward social norms, and less tolerant of frustration than nondrug users. However, no single personality profile or constellation applies to all abusers of all drugs. A possible biological basis for drug addiction is that certain individuals seem to have a more intense response to their first try, which could involve a greater sense of euphoria or a more profound reduction in unpleasant feelings such as anger, depression, or anxiety. The same things hold true, I believe, for people who become intoxicated with or addicted to romance.

It is an old belief that someone using a drug like heroin on a daily basis stops feeling the pleasant effects and continues to use it habitually simply to avoid the unpleasant experience of withdrawal. However, recent studies have shown that even chronic narcotic users still experience a brief period of euphoria following intravenous administration. Moreover, we have a biological model for this, which involves the drug's effects on an animal's threshold for brain self-stimulation.

As discussed in Chapter 2, animals who have electrodes placed in certain areas of their brains will continually stimulate themselves, strongly suggesting that each electrical impulse is producing a pleasant effect. The weakest possible electrical pulse that will cause an animal to continue to hit the self-stimulation lever is called the threshold for self-stimulation for that animal with that particular electrode placement. Various psychoactive drugs such as heroin or cocaine can be shown to lower an animal's threshold for self-stimulation, which means that when given one of these drugs in certain doses, an animal will begin to seek repetitions of low-level electrical pulses that he or she previously ignored. Moreover, drugs that are not abused by humans tend not to have this lowering effect on animals' self-stimulation threshold.

This suggests that drugs to which people become addicted act by lowering the human brain's pleasure center threshold for stimulation, making it easier for the brain areas to be stimulated. This would explain at least some of their euphoric effects, since more external and internal stimuli do feel pleasurable after one of these drugs is taken. Interestingly, while rats become tolerant to many narcotics effects, these drugs do not lose their ability to make the rat's pleasure center more accessible to self-stimulation, even with chronic use. This again parallels human experience, since even long-term heroin users continue to report at least brief euphoric feelings following drug use even though they become tolerant to many of the other narcotic effects. While we discuss this issue in detail in subsequent sections, it seems clear that certain nondrug activities to which human beings appear addicted, such as romance, feel so good in part because they make it easier for our brain pleasure centers to be stimulated.

It is important to note several nonspecific factors that influence how a drug will affect us. The setting in which a drug is taken and the expectations we have of the experience have been shown to influence greatly specific drug experiences. However, again, these should not automatically be viewed as purely psychological issues, since settings and previous experience or knowledge may affect the readiness of certain brain areas to fire, in turn influencing drug effects. Taking a drug in a setting where someone habitually uses that drug seems to enhance its effect.

For any potentially abused drug, the number of users is largely shaped by the drug's availability, and, to a lesser extent, by public attitude. Narcotic use prior to the turn of the century and in certain subcultures today flourishes because the drugs are available and socially sanctioned. The same pertains to amphetamine use in the 1950s; marijuana and psychedelics in the 1960s; and cocaine in the 1970s and 1980s. Availability and public attitude also have important ramifications for the subject of romance, as we will come to see.

DIFFERENT KINDS OF HIGHS

Drugs do not create any new biological reactions, but only alter the rate at which ongoing bodily functions proceed. Therefore, the same biochemical mechanisms that govern drug effects may well play a role in shaping our biological and thus emotional reactions to potent nondrug experience.

Broadly speaking, there are a limited number of different ways of feeling good. Something can make us feel very energized and excited — the thrill that comes from throwing oneself totally into a difficult endeavor or the anticipation of an exciting moment. Or we can feel very happy, euphoric, optimistic — the feeling of "walking on air" when a performance goes well, a long-awaited goal is achieved or appears within reach, or when the future simply seems full of exciting possibilities and empty of threats. Another pattern is to feel intensely calm, peaceful, and relaxed — the experience of "not having a care in the world," or the lightness that comes from "getting away from it all." Yet another is the awestruck, almost mystical or spiritual feelings that accompany something new and overpowering — a vista of rare beauty, an especially intense or intimate moment with another person, or other transcendant experiences of some kind. And finally, there is the relief and excitement that comes when a depression lifts or we come out of a slump, when energy, enthusiasm, and hopefulness return and deadened senses come alive.

Interestingly, but not surprisingly, these same emotional patterns are seen with the various drugs human beings use to get high or improve their moods, suggesting that drug and nondrug highs operate via similar changes in our brain chemistry. These drugs fall into a number of distinct classes, including stimulants, such as amphetamine and cocaine; narcotics such as heroin, opium and morphine; antianxiety agents such as Librium and Valium; sedatives such as barbiturates, Quaaludes and other "downs"; alcohol, which chemically acts much like the sedatives; marijuana and other cannabis derivatives; and psychedelics, such as LSD, mescaline and psilocybin. Other classes of drugs, such as antipanic agents and antidepressants, do not produce positive feelings in most human beings, but have mood-enhancing effects in people whose brain regulatory mechanisms are abnormal to begin with. As we look briefly at each class of drugs, keep in mind that we are searching for clues as to how our brains operate in situations of pleasurable stimulation.

A stimulant high is usually happy and energetic. Within thirty to sixty minutes after taking 10mg of amphetamine by mouth, one sees increased alertness, decreased fatigue, elevated mood, greater initiative, and an enhanced sense of self-confidence. This is often accompanied by feelings of elation and euphoria, as well as more physical activity and talking. Simple tasks are easier to do, athletic performance can be enhanced, and more work done, although not necessarily with fewer errors. Intravenous effects tend to be more rapid and more intense. Prolonged use can be followed by mental depression and fatigue when the drug is stopped, or by paranoia leading to psychosis if continuously administered over a period of days or repeatedly used in escalating doses over a longer time.

Most human beings find single doses of an amphetamine to be generally pleasurable. Interestingly, however, certain depressed individuals, whose brain chemistry is disrupted in some way, do not become even temporarily euphoric in response to amphetamine and sometimes actually feel more sad or despondent following drug administration.

The purpose of this discussion is not to get people interested in taking amphetamines. What these drugs do, however, is show that we all carry around within us a capacity for energy and excitement that rarely gets fully utilized. What stimulant drugs (this includes cocaine, which is pharmacologically quite similar to amphetamines) do is tap into a brain chemical "reservoir," causing an outpouring of certain neurotransmitters, which throw our systems into high gear. Given the similarities between drug-induced and naturally occurring excitement, it would appear that any nondrug experience that makes us feel more lively, interested, and enthusiastic does so by tapping into this same brain chemical reservoir. Just because the chemical effects are the same, however, in no way means that amphetamines are a valid substitute for pursuing natural pleasure or excitement. Our nervous systems have evolved so that we will seek certain kinds of romantic and other rewarding experiences. What stimulant drug abusers do is short-circuit this time-tested process by inducing a sensation of excitement without our having to accomplish anything.

This capacity to feel energized and excited is the mechanism that first makes us feel we are in love. Falling in love is having your pleasure center go bonkers in response to an interaction with another person. One way to test this would be to slip people a stimulant just as you introduced them to someone of the opposite sex. This would actually be a variation on the Schacter-Singer experiment described in Chapter 1. My prediction would be that if you felt very turned on and thought it was because of the person you'd just met, you might temporarily feel you'd just fallen in love. The problem would be that, unlike the real thing, when the drug effect wore off you'd just as quickly fall out of love.

While amphetamine and cocaine highs are clearly related to the stimulant effects of the drugs, the case is not so clear for the narcotics, a class of medically used painkillers that includes heroin, opium, morphine, and codeine, among others. The problem is that unlike amphetamines, narcotics also lower anxiety, which may partially or even entirely account for the euphoria they induce.

We are only beginning to understand the relation of narcotics to anxiety. Narcotics suppress the activity of a particular brain area that may act as an anxiety center. This area (the locus ceruleus, or LC) appears to be heavily involved in producing terrifying panic attacks and perhaps even the anxiety we feel on separation from loved ones.

Animal experiments also suggest that narcotics can lessen separation anxiety. These drugs are very effective in reducing the crying of puppies, young guinea pigs, or chicks that are separated from their mothers or peers. Other drugs cannot do this without knocking the animals out, whereas narcotics reduce or abolish crying while not impairing the animals' ability to function. In some experiments narcotic blockers have been shown to make young animals cry more after separation, most likely by blocking the effects of the body's own internal narcotics. These in turn may function to reduce at least partially our distress when we are separated from loved ones.

Given that they produce mild euphoria and also alleviate anxiety, that the type of anxiety relieved may relate to separation from loved ones, and that these effects don't seem to wane with time, narcotics seem like a good model for how a long-term love relationship affects us. When the relationship is going well we feel good, or at least calm, but if something threatens it, be this loss of affection, illness, or some enforced separation, we feel scared and in pain. Moreover, we certainly seem to get "addicted" to our long-term partners, a phenomenon that carries none of the social disapproval of being addicted to a drug. Once again this may be because it is socially useful and adaptive for us to get "hooked" into long-term relationships, while getting "hooked" on a narcotic drug is just the opposite. But more about this in a bit.

Tranquillizers such as Librium and Valium are widely used in our society. Some experts see this as evidence of widespread abuse; others as a sign that the drugs effectively alleviate the anxiety many people feel.

What positive effects does a tranquillizer have? In a variety of animal species, drugs like Valium and Librium increase behavior that has been previously eliminated by punishment or other negative training. The drugs are thought to reduce fear or anticipation of punishment, which are close parallels to human anxiety. Thus, these drugs help us to do things which we have become afraid to do. The activity can be an active one, such as making love or working toward a goal, or a passive one, such as trying to relax, sleep, or simply enjoy something.

We now have some leads on how this happens. Valium and Librium seem to interrupt certain pathways, suggesting the existence of a brain "punishment center" that the drugs turn off.

The situation is yet more complex, however, because of the recent discovery that there are specific receptor sites in human and animal brains for drugs like Valium and Librium.

What role could these naturally occuring Valium or Librium receptors possibly play? As with the narcotics receptors, the first thought is that there must be naturally occurring chemicals in our bodies that fit these receptors. It turns out that there are, but their actual role so far is not well understood, and in fact, is a matter of controversy. One school of thought holds that our bodies make their own tranquillizers, small molecules that fit our Valium receptors and help calm us down in times of stress. An opposing view is that there are naturally occurring blockers that fit these receptors, and that these increase rather than decrease our feelings of alarm. In situations of real danger or threat this could be very useful, but in many people who suffer unnecessary anxiety the system may be overactive.

Our anxiety systems may well play a role in how we approach others romantically. Most all of us have been turned down at least a few times when asking someone for a date. How is it that we (or at least most of us) don't become fearful and avoidant of trying again? And why, in fact, do some people actually become so traumatized that they stop trying? While we are far from knowing the complete answer to this question, perhaps our brain tranquillizer systems get called in to play in the face of a rebuff or negative experience, either in terms of causing us to feel the fear (if the alarm system theory is correct) or in reducing our distress so we are willing to try again tomorrow (if the internal tranquillizer people are right). Finally, as for why some people are more easily traumatized by failures than others, perhaps this is due to inherited differences in our Valium and Librium "key" and "lock" systems.

Alcohol is grouped with sedative drugs like barbiturates and methaqualone (Quaalude) because, pharmacologically, their actions lower the activity of our nervous systems. The high they produce is generally thought to be due to the fact that what gets shut down first are such things as rational judgment and self-restraint. What results is a "liberation" of more primitive brain and emotional activity, resulting in increased confidence, gregariousness, dramatic mood swings and dramatic outbursts. With more widespread brain slowing, we get sleepy, but feel very relaxed and calm. Given alcohol's popularity in our society, it is reasonable to conclude that Western man finds pleasure from time to time in shaking loose from certain "civilized" restraints and allowing more "primitive" impulses freer rein. As we will see, this disinhibition model is quite descriptive of a variety of exciting activities.

Romance also has a disinhibiting effect for many people, which is one of its appeals. In general lovers are prone to feel bolder, more gregarious, less embarrassed, and more oblivious to what is going on around them. What I think is involved here is a disruption of the normal balance between social restraint and primitive emotional drive, similar actually to what goes on with alcohol use. The difference is that with alcohol or sedatives, inhibitions and social restraints are put to sleep, whereas with romance, our emotional drives simply become too powerful to keep totally in check. And for those lovers who can't quite say what is on their minds (or in their hearts) a drink or two is often quite helpful.

Where the alcohol model fits romance (used in the loosest sense) best is in wild swinging types whose idea of a good time is a sexual orgy. Human society has always had this orgiastic or "Dionysian" streak which traditionally emerged in wild and frenetic feasts, dances, and other rites. We have moved away from such things as we have become more "civilized," which is why alcohol still has such a powerful appeal.

The feature that distinguishes the psychedelic drugs (LSD, mescaline, psilocybin) from the other classes of drugs is their capacity to alter our perceptions, thoughts, and feelings in ways that we otherwise experience only in dreams or mystical states.

Initially I was unsure as to whether there was any connection between psychedelic and romantic experiences. The link became clear when I recalled Abraham Maslow's description of "peak experiences," which are brief moments in your life when you may have felt totally immersed in something, immensely fulfilled by it, and awestruck by a sense of beauty, meaningfulness, and timelessness. This description has often been applied by drug aficionados to the "best" of their psychedelic experiences, by others to moments of intense creativity, or perception of great beauty by seekers of mystical religious experiences, and *by lovers* at times when they felt totally at one with their partners. Is this all just coincidence? Or could it be that all these experiences tap into the same brain chemical systems?

The types of drugs discussed up to now exert their effects in just about anyone who takes them, suggesting they interact with our normal brain chemistry. Such is not the case, however, for medications that block or reverse severe or chronic depressions. Rather, these drugs, which are profoundly helpful to people with certain kinds of disturbances, have little or no effect in normally functioning people. What this suggests is that in states of extreme depression our normal emotional control mechanisms are out of kilter, and that antidepressants act by normalizing these brain systems.

Earlier I described Mr. T., whose pleasure center had completely shut down, a condition we call melancholia. There is another pattern of depression where people lose their ability to anticipate pleasure, but can still enjoy certain things. Left to themselves they prefer to sit and do nothing, because they don't think doing anything (going to a party or a movie, for example) will be any fun. Yet if they force themselves to go, or if someone drags them along, they will actually have a good time. This particular pattern of pleasure center disturbance is associated with a syndrome we call atypical depression and can often be helped by a different kind of antidepressant than is useful for melancholia.

Clearly there are at least certain types of depressions which are related to pleasure center malfunctions. Interestingly, while we usually think of people as having depressed periods, some actually seem always to be depressed, or at least not to get much fun out of life. This takes us to why people differ in their capacities for pleasure.

WHY PEOPLE DIFFER IN PLEASURE CAPACITIES

Why do some people seem to lead energetic and vital lives while others find the world uninteresting and tedious? To some extent how much fun we have depends on our life circumstances, health, and so forth. Part of the equation is psychological and reflects such things as early conditioning, parental and subcultural sanctions. But what if there are also biological influences that shape our pleasure capacities and influence what kinds of experiences we enjoy and how much we enjoy them?

One area that this is undoubtedly true is for chronic, low-grade depression, which is often chalked up to personality problems or an existential view that sees life without meaning unless we struggle to give it one. Actually, I've often wondered if some of the existential writers who struggled so to find meaning in life were chronically depressed, because barring catastrophic life experiences, people with normally functioning pleasure centers seem to find some joy and purpose in life without having to search for it. What a number of recent research studies have highlighted is that many people with chronic or lifelong low mood states feel much better after taking antidepressant medication. Since these drugs do not provide a mood lift to nondepressed people, this suggests that people who are helped have underlying biochemical disturbances.

Other attempts to explore the biological basis of our pleasure capacities are in their earliest phases, and work in this area must be considered preliminary at best. Yet a number of exciting projects are going on, and leads have been developed. To round out our understanding of our human ability to feel high, these will now be considered.

SENSATION SEEKERS

Certain types of people seem to thrive on and even require massive doses of new stimuli to feel good and excited about their lives. These people have been labeled "sensation seekers" by the psychologist Marvin Zuckerman, who has designed a questionnaire to identify them. Zuckerman distinguishes four varieties of sensation seekers — the "Thrill and Adventure" types who like outdoor sports or other activities involving danger and speed; the "Experience Seekers" who pursue many forms of inner experience through travel, marijuana and hallucinatory drugs, modern music and art, and an unconventional life-style; the "Disinhibition" types (the "swingers"), who like wild parties, sexual variety, gambling, and heavy social drinking; and those people who score high in "Boredom Susceptibility" and can't stand routine, repetition, monotony, or dull people. Interestingly, sensation seekers tend to rate low on psychological tests for anxiety and low on needs for nurturance and orderliness.

Society has always debated how to view people very high in sensation seeking. A positive view is that they are less bound by anxiety to conventional pursuits and inhibitions, and therefore freer to pursue some of the many pleasures and adventures life has to offer. A negative view is that they are more impulsive, restless, or maladjusted, requiring continuous novelty to substitute for their lack of commitment or to ward off a tendency toward depression. Either view is an overgeneralization, and individuals high in sensation seeking, as is true for most aspects of human behavior, vary in how healthy and neurotic they are.

Traditional views also ignore what may be the biological aspects of sensation seeking, both in terms of how much novelty a person seeks and what specific pattern of stimulation he or she prefers. For example, the thrill seekers seem to derive maximum pleasure from movement and changes in body position, as well as from high levels of arousal. It has been proposed that there are pathways running from the area of the brain that regulates and coordinates movement (our cerebellum) to the limbic system, which could be why we find gentle rocking pleasant. Thrill seekers could differ in some way in regard to this pathway.

Experience seekers, on the other hand, like more modest arousal, primarily of their mind and all their senses. Using our drug model, one could also speculate that the brains of people high in boredom susceptibility become tolerant very quickly to situations or other stimuli. Most interesting of all, the brains of people high on the disinhibition scale seem to handle increasing stimulation differently from people who score low on that scale.

In a test measuring average evoked response (AER), subjects are hooked up to an electroencephalograph (EEG). They are then exposed to brief flashing lights of different intensities. An AER augmenter is someone who shows increasingly large brain wave response when light intensity increases, while an AER reducer records progressively smaller brain waves. The nervous system of an augmenter is supposedly able to continue to fully process stimuli as they get stronger, while a reducer's brain begins to inhibit or tone down its reaction to the highest levels of stimulation.

What the research suggests is that people who prefer uninhibited sensation seeking have nervous systems with higher
tolerance for intense stimulation than do people who score lower,
which might explain their different patterns of behavior.

Within the sensation seeking group as a whole, the uninhibited
types seem to favor alcohol, with its loosening effect in social
constraints, over drugs, while the experience seekers prefer the
psychic effects of marijuana and the psychedelics. These two
types also seem to differ in their approaches to romance, since
the disinhibited folks like casual sex and avoid commitment,
while the experience seekers emphasize commitment and depth
in their relationships. As we will see, the different preferences
of "swingers" and "lovers" have been noted many times before,
but no one has even stopped to question whether there might
be a biological explanation.

MANIA

No discussion of feeling high would be complete without the
mention of hypomania and mania. These are psychiatric terms
that refer to a range of feeling and behavioral states which have
in common an acceleration of mental processes and physical
actions. Hypomania is a mild to moderately overexcited or "up"
state in which someone seems more optimistic or happy than
circumstances warrant. The hypomanic person needs less sleep
than usual, starts more projects than probably can be completed,
is always on the go, talks too much, and so forth. Mild hypomania can be very functional, providing as it does great optimism,
intense energy, enhanced self-confidence, and so on, as some of
our great salesmen and entrepreneurs can testify. Others think
mild hypomania makes people more perceptive, more imagina-
tive, and more creative.

More severe hypomania, which is still not always recognized as a psychiatric condition, tends to be more of a problem. Too many projects are started and plans made, too much noise created in the middle of the night, too much money spent, too much irritability if someone disagrees with or protests about what is going on. A severely hypomanic person is also highly distractable, so that the sustained concentration needed for productive or creative effort becomes more difficult.

Mania is usually less subtle and therefore more easily recognized, although for years it was often misdiagnosed as schizophrenia. People in full-blown manic episodes are out of control — they don't shut up, they never stop moving, they hardly ever sleep, they make totally unrealistic plans, and they think anyone who disagrees with them is stupid or crazy. In extreme states manics become delusional and completely lose touch with reality, usually in a grandiose way such as thinking they have a divine mission to perform or have discovered a cure for cancer. Manic individuals can also become paranoid, believing, for example, in conspiracies against them or that their cure for cancer is being suppressed.

What is so fascinating about hypomania and mania is that they are episodic, and occur in people who also have periods of depression. Someone will be going along in a normal state and all of a sudden things begin to change — energy increases, need for sleep decreases, horizons broaden, optimism and confidence expand. Then, for reasons we don't at all understand, but which are undoubtedly genetically and biologically determined, some peak in this mildly high state, while others take off into the stratosphere. An episode can be as brief as a day or as long as six months; when it's over, some plunge into depression, while others glide back into normality.

If it weren't for the twin risks of full-blown mania and devastating depression, I would be very envious of the energy and optimism that my hypomanic patients seem to have. What I can summon up on a very good day or more routinely for a few hours, they possess for weeks at a time. The chemical basis is probably similar, but they are not being regulated in the same way, so the normal dips of energy with fatigue or daily fluctuations in mood and optimism don't seem to occur. On the other hand, positive life events including romantic encounters seem to fuel or prolong hypomania, suggesting that they induce the same sort of biochemical changes that cause the hypomanic state in the first place.

The brain mechanisms that actually cause mania or hypomania are poorly understood. That mania and hypomania are due to altered brain chemistry is demonstrated by the fact that lithium carbonate, a simple salt, can reverse and prevent the dramatic alterations of these states. Lithium has much less effect on normal human mood or functioning, although it has been reported to increase fatigue and decrease sense of well-being. Prior treatment with lithium also blocks much of the euphoric and activating effects of amphetamine and cocaine as well, suggesting that stimulant effects and mania have a similar biochemical basis. Lithium has also been shown to reduce self-stimulation in rats. Lithium's ability to block mania or drug stimulation is probably due, at least in part, to a normalizing effect on brain pleasure centers.

But we are all to some degree physiologically and biochemically unstable, which is what allows us to experience romantic ups and downs as well other shifts in mood and energy. It is thus tempting to say that manic depressives are simply at one end of a spectrum of biochemical instability, but that all of us lie somewhere on this continuum.

What people who become manic (or severely depressed) lack, however, is some built-in stabilizer or chemical thermostat that keeps their brain chemistry from shifting too far in any one direction. While certain shifts are useful and allow us to feel excited or even exalted at times as well as appropriately sad, changes in brain function that go too far in any particular direction can render us totally unable to function.

The key question is why some people have brain pleasure circuits that permit them to live with zest and vigor, while others either swing too widely, or not widely enough. While we are nowhere near being able to answer this question fully, let me close this chapter by describing a research program that at least has focused on it.

THE MAO STORY

Most of you probably associate MAO with the former leader of mainland China. But to present-day psychiatrists and brain biologists, MAO has another meaning. It stands for monoamine oxidase, perhaps the most important brain enzyme in the regulation of emotional states.

An enzyme is a chemical substance that causes a change in another substance. The digestive enzymes in the stomach and small intestine that help break down food are familiar examples. Monoamine oxidase is an enzyme, or really a class of at least two enzymes, that helps the brain and body break down (metabolize) certain simple chemical substances, called monoamines, by a chemical process called oxidation. What makes monoamine oxidase or MAO so important is that this enzyme class has, in biological terms, a particularly ritzy clientele, since it happens to be a major metabolizer of several neurotransmitters (norepinephrine, dopamine, serotonin) and also phenylethylamine (PEA).

What this means in practical terms is that our bodies rely on MAO to help regulate the amount of several important transmitters available for brain activity at any given moment. The greater the MAO activity, the faster the breakdown of these substances and the less available they are. Conversely, the lower the MAO activity, the greater the level of brain monoamines.

This principle lies behind psychiatry's first true antidepressant, a drug called iproniazid, which blocks the action of monoamine oxidase and therefore is called a MAO inhibitor. The finding was accidental. Iproniazid was developed as a new drug to treat tuberculosis because of its similarity to the well-established anti-TB drug isoniazid (INH). When given to patients on a chronic TB ward, however, iproniazid seemed to lift the moods of many of the justifiably depressed patients more than could be accounted for by improvement in their tuberculosis. Some smart clinicians then began to use iproniazid as an antidepressant.

The fact that first iproniazid, and subsequently a whole class of MAO inhibitor drugs, have proven to be highly potent antidepressants has been one strong pillar for the theory that depression is caused by a decrease in level or activity of one or more neurotransmitters. That MAO inhibitors can throw certain vulnerable people into mania also buttresses the notion that excess brain chemical activity is the basis of mania.

If MAO levels are important, then the next question is: Do people with different emotional profiles differ in the levels of MAO? Psychiatrists have been particularly interested in this question in their search for the causes of depression, mania, excessive anxiety, and even schizophrenia. All the enzymes that contribute to the formation as well as breakdown of neurotransmitters have been studied. Interestingly, the only one that has turned up abnormal in a number of studies has been MAO.

The MAO studies that I am about to describe involve measurement of the enzyme in blood platelets, which seem to reflect what goes on in the brain. All were conducted at one of American psychiatry's foremost research centers, the National Institute of Mental Health (NIMH) in Bethesda, Maryland. This federally supported research complex has played a principal role in training psychiatric researchers and financing their activities.

One study published in 1977 found that young men with lower MAO tended to have higher sensation-seeking scores, with a preference for wild or disinhibited forms of behavior and a particular distaste for anything that smacked of the routine or felt tedious or boring. Interestingly, this finding did not hold up for the females in the study.

The next study, reported in 1978, also found that in male community college students, lower MAO went along with higher scores on all the sensation-seeking scales. Women with lower MAO had higher sensation-seeking scores in general.

Low MAO was also associated with a number of emotional disturbances in the student population, including difficulties that resulted in psychiatric hospitalization, suicide attempts, and criminal behavior. The overall thrust of these findings is that low MAO is associated with several types of extreme behavior, some good, some bad. Perhaps low MAO simply makes extreme behavior of some types more likely, with the exact nature of that behavior being shaped by other biological or psychological factors.

How people spend their leisure time was also examined in these studies, and an association was found between low MAO — which should mean more neurotransmitter activity — and a greater tendency to seek stimulation at rock concerts, museums, and plays. Interestingly, a study of MAO in rhesus monkeys found that those with lower levels were more social, more active, more playful, and slept less. Thus it would appear that some low MAO people have more fun, while others get into more trouble.

When we talk about looking for biological (or other) influences of normal human behavior, however, we come up against a strong bias which takes the form of "I don't really want to know what makes me tick." This was exemplified by the statement Senator William Proxmire made several years ago when he gave his Golden Fleece Award for wasting public funds to a team of psychologists who were studying romantic behavior. The essence of the feeling seems to be that efforts to understand human emotions or behavior in areas such as romance are frivolous at best, and could even spoil the poetry and mystery of our lives.

Rather than join in the debate I would like to let science and the spirit of humane scientific inquiry speak for themselves. For we are about to jump into the heart of the controversy as we now turn to an examination of the biological underpinnings of romantic feelings and behavior. As we have seen thus far, extreme biological flux, whether drug induced, experientially caused, or even unprecipitated, can lead to patterns or states of exhilaration on one hand and profound misery and intense emotional pain on the other. Interestingly, but not surprisingly, these same terms aptly describe the range of feelings we go through during romantic relationships.

CHAPTER 4

ATTRACTIONS AND ATTACHMENTS:
Our Romantic Wiring

A thirty-five-year-old woman named Jennifer came to see me several years ago. She was an advertising executive who had been depressed for much of the past two years, since ending a six-year marriage. During the marriage her mood, energy, and feelings of self-worth, not to mention her ability to perform on her job, had fluctuated dramatically. She felt well and did well when the relationship was going nicely, and did very badly when it wasn't. At such times she would feel very depressed, would overeat and gain weight, not see her friends, spend most of her time in bed, and miss work or do very badly at it if she could force herself to go. On the other hand, when things were going well with her husband, she was energetic, capable, bright, and lively. Many years of psychotherapy had not altered this pattern at all, and one attempt to take the usual type of antidepressant had not proven helpful.

Since ending the marriage she had had several brief romances. These cheered her up temporarily, but the good feelings didn't last because none of the men was available most of the time —

which Jennifer experienced as devastating personal rejections. Thus for most of the two years before I saw her this woman felt miserable. She gained twenty pounds, dropped most of her friends, lost several jobs, and spent most of her free time in bed.

Putting her on a specific type of antidepressant has dramatically changed her life. Her mood lifted in several weeks, and has remained good for the past two years. She lost the twenty pounds, became active with her friends again, and has done extremely well in her professional life. *Most importantly, she has become able to feel positive about herself and her life even when she is not involved with a man*. She is still very excited by a man's interest, and in fact continues to see several men friends from time to time. The great change is that she does not crash into depression when they are cool or aloof or unavailable. Now she can carry on whatever happens, making her an independent person for the first time in her life.

What is so interesting about this patient is not that an antidepressant improved her mood, which after all is what the drug is supposed to do. Rather, something else changed, namely her feelings about herself and about her life. Prior to receiving treatment she had had periods of good mood and periods of low mood, but always there was a need for a man to make her feel worthwhile and adequate as an individual. This no longer seemed to be the case after treatment. So for the past two years she has felt okay whether or not she had a romantic partner.

This pattern is by no means uncommon. The key seems to be an extreme sensitivity to feeling rejected, usually leading to recurrent depressive crashes after romantic disappointment, coupled with an intense need for romantic involvement to maintain normal mood and self-esteem. While some people like Jennifer get into prolonged slumps, others seem to live on emotional roller coasters, as they oscillate between extremes of elation and despair, depending on the state of their current relationship. Surprisingly, many individuals with this problem are not losers in any conventional sense. More often women than men, they can be young, bright, attractive, and talented. Moreover, they are almost all very engaging and stimulating. Most have no trouble getting people to become interested in them. Rather, their problem lies in being unable to form lasting relationships, both because of the poor choice of partners (often aloof, married, etc.) or because the continued need for attention, praise, and reassurance often drives the other person away. These factors, coupled with the need to be in a romantic relationship to feel and function well, make life very difficult.

One might be tempted to say at this point: "So, what's new?" People like this have been described before — actors or actresses who need constant admiration to feel good about themselves, clinging or "insanely" jealous lovers who suffocate their partners and eventually drive them away.

What's new is that these particular behavior patterns appear to have an underlying biochemical aspect — that is, they may stem not exclusively or even primarily from problems with people's psyches, but from defects in their body chemistry. What we have found in our work with these patients is that putting them on certain types of antidepressant medication, as in Jennifer's case, has dramatic effects, not only in lifting their moods, but in markedly altering the way they interact with others. Aside from being helpful for people like Jennifer, this work is making us take a whole new, chemical view of what goes on inside the human body and brain when we are in (or out) of love.

In speaking of romantic love, I am referring to a very special sort of human feeling and bond, one that has not been common in all cultures or in all periods of human history. While hard to define precisely, romantic love is a set of feelings that, all together, make us feel as though we're in a unique emotional state.

These feelings include a sense of intense excitement, great calm or greatly enhanced well-being in the presence of the other; a desire to be alone with, to reveal oneself to, and be known and understood by the other; a strong desire for sexual intimacy, whether acted upon or not; possessiveness in regard to attention and affection from the other; a strong concern for the welfare of the other; a sense that one's life would be greatly diminished by the loss of the other; and an element of idealization, which involves seeing the other as more attractive, noble, intelligent, or otherwise gifted than he or she may actually be.

Not every one of these feelings must be present to call a relationship romantic. For example, youngsters of nine or ten and "oldsters" in their eighties develop what are considered to be romantic relationships that seem on the surface to lack strong sexual desire. Yet the hitting and teasing in pre-adolescents who have crushes on each other are probably related to sexual impulses, while sexuality in the elderly has been traditionally limited both by cultural prejudice and medical disability. So even in these cases, sexual attraction is most likely part of the whole picture, and helps distinguish romantic relationships from other strong emotional bonds. However, the distinctions between our romantic and other close relationships are not so absolute, so some of what is said about romantic states will also apply to our other relationships as well.

Interest in romantic love is not new in Western society. Yet it is fair to say that our society today is preoccupied with romance as never before. Today, romance is a major focus of life, with enormous amounts of time and energy devoted to meeting, dating, mating, and staying with suitable partners. Unfortunately, for reasons that seem to be at least partially biological, more and more time and energy are also devoted these days to separating from no longer suitable romantic partners.

Given all this, it might seem logical to ask: Why is this happening? Why has romance attained such a central importance in our lives? Have we been brainwashed by the media? Or do we as a society have the time and freedom for romance not available to earlier generations or to peoples of other cultures? Is romantic love as we know it some sort of cultural craze, or an expression of our human potential?

But first we must ask even more basic questions, namely what romantic love really is, and why so many people pursue it so vigorously. Once we've answered these, perhaps then we can tease apart those aspects that seem biologically wired in from those which are taught to us.

Most studies of human behavior and motivation have neglected to include the biological side of man's behavior. Nowhere is this more clear than in love and romance, the study of which has relied for the most part on surveys by social psychologists or case studies by psychoanalysts. While both have made contributions, it's time to look at the biological aspects of romantic feelings and drives.

THE BIOLOGICAL BASIS OF ROMANTIC FEELINGS

Romantic love as we know it is an intense emotional state that one person experiences in relation to another. While the cause for this state seems to be another person, there are other processes going on inside of us. Since the way we experience emotions in general is rooted in those neurochemical circuits of our brain's limbic center that we discussed earlier, any intense emotional experience requires that some pattern of brain activity be kicked off. The question then becomes: Do the heady emotions of romantic love involve special or unique changes in our brain chemistry or do they overlap with what goes on within us during other kinds of intense emotional experience?

As you will see, what is unique about romantic love is not the kinds or amount of emotion we feel at a given moment, but rather, what triggers these emotions. Romantic love involves several intense forms of brain arousal triggered by contact with or thoughts about some specific other person. These kinds of arousal appear to be among the most intense that most of us are capable of, and at least in some cases, grow rather than diminish with time. But in its biochemical patterns, romance appears to be similar to a variety of other intense human experiences, suggesting that the human nervous system may have only a limited number of ways of responding.

For primitive man two aspects of relating to the opposite sex were important for survival as a species. The first was to have males and females become attracted to each other for long enough to have sex and reproduce. The second was for the males to become strongly attached to the females, so that they stayed around while the females were raising their young, and helped to gather food, find shelter, fight off marauders, and teach the kids certain skills.

These two sets of feelings are the basis of modern relationships as well. Biologically, it appears that we have evolved two distinct chemical systems for romance; one basically serves to bring people together and the other to keep them together. The first is attraction. Attraction is the excitement we feel when falling in love, and is quite similar to what happens when we take a stimulant drug. The second, which helps keep people together, is attachment. Attachment has more to do with feelings of security than of excitement. It too has a number of drug analogies, although surprisingly it may have more to do with narcotics than any other drug type.

While at first it seems strange to compare romantic feelings with drug states, it makes more sense if you remember that drugs do not create new chemical reactions in our bodies, but rather, speed up or slow down existing processes or interact with existing receptors. If these same chemical reactions or receptors are involved with our romantic passions, then romantic states are going to have much more in common with drug states than is usually thought to be the case.

ROMANTIC ATTRACTION: HOW TO GET THEIR ATTENTION

Attraction is what you feel when you've just met someone who really excites you and you begin to fantasize about where you might see him (or her) again. This is your basic romantic rush, the churning in your gut when you've met someone who really turns you on; your eyes meet and the other person seems to be responding well. This may sound like sexual arousal and in some ways it is, because sexual attraction may be the start of romantic attraction, at least for many people. But romantic feelings, typically colored by idealization and the first whispers of attachment, soon extend beyond sexual passion.

Strong attraction, especially to someone with whom you might become romantically involved and who also seems attracted to you, is an extremely powerful emotional state. You feel excited, even elated every time you think about seeing the person you've met. Your sense of self changes — you feel more attractive, more confident, more capable. You feel more optimistic about the future. Your body feels different — more bouncy, more energetic, and in need of less food and sleep. You may start to view the person you've met in glowing, larger-than-life terms — the most wonderful, the best looking, so sensitive, incredibly intelligent, etc. What's more, two people who have fallen in love amplify these feelings for each other by the signaling that goes on between them: the looks and gazes, the body language, the excited wish to know and be known, and the warm acceptance of what the other has to say.

There are many similarities between this emotional state and what happens if you take a stimulant drug. What amphetamine does for most people is rev up their motors and lower their pleasure thresholds. Their energy goes up and their need for sleep and food goes down. Everything is seen in a more favorable and exciting light — especially the person's own abilities and attributes, the abilities and attributes of others, and what the future has to hold.

Our capacity for looking forward to things is like this amphetamine response. When we are looking forward to things, especially when we are pursuing a valued goal, we liven up, have more energy, and concentrate better. It is the excitement of the chase, the thrill of keen competition — it sharpens your senses, whets your appetite — you can almost taste a victory that you want very much.

Strong attraction, especially when there is a sexual component, taps into the same neurochemical circuits as do other exciting or highly stimulating pursuits. This is really not so surprising because when we feel attracted we often begin a pursuit, one whose outcome is uncertain and is filled with dreams and fantasies (anticipations) about the future. The prize is some imagined state of incredible happiness, be it sexual bliss or a lifetime together. But the threats or obstacles to be overcome are many. Either of you could cease to feel attracted; one of you could develop a fatal illness or be killed in a car crash; or the two of you could simply be unable to "get it together."

If strong attraction involves the same neurochemical circuits that are set aglow by the stimulant drugs, this could account for the bad judgment that both amphetamine users and lovers sometimes show. When we view something in an unrealistically positive way, we are headed for trouble. It might involve the feeling that "there is nothing to worry about" when in fact there is plenty to worry about. Or it might involve ignoring the shadier aspects of a newfound partner's personality or past because we are so strongly attracted. What is going on in either case is something that psychologists call idealization, which means seeing things, usually another person, in unrealistically positive or idealized terms. At the same time we tend to deny difficulties and minimize or make very much less of things that we don't want to deal with.

If feeling attracted is biologically like taking amphetamine, then the inability to see your partner realistically in the early phases of a heady romance may have a neurochemical aspect to it. As long as the new relationship is going well, it is as though our brain pleasure center threshold is turned way down so that many things can reach it, and your displeasure or alarm circuit thresholds turned way up so little gets through. But this state is delicately balanced, because if something starts to go wrong in the relationship, the whole thing begins to fall apart. A tiff with your new partner, and your mood can plummet. A real breakup, and you can be plunged into despair. At such times your brain's limbic centers seem to do a rapid flip-flop, and now nothing can register as pleasurable.

How long strong attraction takes to develop seems to vary. For many people it happens very quickly, even on a first encounter. Sometimes it may even involve seeing someone from a distance and feeling that you are suddenly smitten. For others it takes somewhat longer, and requires time to get to know the other person. But attraction is usually hasty, a powerful emotional force that is kindled quickly when we meet someone who seems right.

Not too long ago a young woman came to see me. Her story illustrates a lot of these points. Bright, charming, and attractive, she had had a series of relationships that had all ended unhappily. Some of the men had left her and each time she had then crashed into depression. But others had remained interested; in each of these she herself gradually lost interest.

The traditional psychiatric approach would view this woman as having an unconscious need to fail romantically, so that one way or another she avoids having to make a commitment. But let me continue with the story.

What turned out to be so interesting for me was that Ms. X saw romance as a "quick fix," something that could be counted on to make her feel happy and elated (at least in the beginning). She would meet someone and if the "chemistry" was right, she'd fall in love. Then, at least for a while, she'd feel terrific.

But why didn't it ever last? I think there were several reasons. One is that this woman fell in love so quickly she usually didn't take a good look at whom she was falling for. Some were married, some committed to remaining unattached, some still hung up on a previous love. By the time the full story would come out, she was already hooked. But also these men represented a special challenge — to win them over against the odds. And despite her bravado, she wasn't self-confident enough to really turn one of these types around.

Then why dump the guys who liked her? Well, again you could say it was fear of success. Or could it be that with these men things simply got boring; that once they'd sufficiently professed their love they offered no further challenge, no further pursuit? What we are dealing with here is a lady who gets a real amphetaminelike high from the opening phases of romance and hasn't yet been able to get beyond this.

IS ROMANTIC ATTRACTION THE SAME AS LOVE?

Many people believe that you can fall in love with someone you see from across the room even though you've never actually met them. This is part of the "love at first sight" debate. Some psychologists believe that love at first sight is arousal, not love, and that when it happens you are being turned on by a fantasy, not falling in love with a person. People who seem to fall in love so easily are thought to carry around in their heads an idealized image of the kind of person they want to be in love with, one who will presumably meet all their important emotional needs. They walk into a room, see someone well cast for the part, and whoosh, all the dreams and feelings that surround that fantasy are suddenly transferred to a real person, with whom they are then madly in love.

My own view is that attraction, no matter how powerful, is by itself not the same as love. Romantic love requires attraction plus attachment. Someone can be very strongly attracted to another person but never become attached. Without some feeling of attachment, attraction is nonspecific. You can be very attracted to someone one minute, and lose interest completely the next if someone even more attractive walks by. But once you have begun to fall in love with someone, this person becomes special for you. This process by which we get "hooked" on someone, or more accurately, hooked onto them, is the process of attachment.

Does this do away with the notion of love at first sight? It all depends on whether you feel attached, as opposed to attracted, to another person very quickly. If it's only attraction, then I would call it arousal, not love. But if you feel actual attachment, even if it's to a totally fantasized image of another person, then it's love.

Interestingly enough, attraction and attachment appear to be neurochemically different although often overlapping processes. Someday it may be possible actually to monitor chemically what goes on between two people and decide when attraction leaves off and attachment begins. But until we can begin actually to study the process biologically, it is a matter of semantics as to how long it takes for true love to develop. When you think about it, it makes very little sense that we are capable of developing such strong emotional feelings, be it attraction, attachment, or love, for people we don't know very well at all, which is true for someone we fall in love with after six months as well as six minutes. All speeds of passionate falling in love involve some idealizing of the partner, because that is what this type of love is all about. There are very few real gods and goddesses around; the rest of us rely on being dressed up at least to some degree by our partner's fantasies and idealization. One could argue that the longer we know someone the more time we have to learn about what really makes them special, and I would agree. But this keeps lovers together. What brings them together is always something of a fantasy or a dream — here is the specific person I've been waiting for, the one with whom I will live happily ever after.

THE CHEMISTRY OF ATTRACTION

It is interesting to speculate on the actual neurochemical basis for our feelings of attraction. In likening it to a stimulant drug state, I believe I am doing more than making an analogy. What seems likely is that the same neurochemical events that underlie many kinds of pleasure and stimulant drug arousal are also involved when we feel very attracted to someone. What we are talking about then is some kind of increased chemical activity, involving either increased neurotransmitter levels or increased receptor sensitivity.

That sexual and romantic attraction was chemical at all was brought home to me in the course of working with a number of patients. One was a young man who found romance particularly intoxicating, and used to tell me that "falling in love was like taking amphetamine." On one occasion he met someone he really liked, and the two of them spent the next five days together. What made this a little unusual was that they barely slept during that time, and also never spent more than one day in the same city. They met in New York, went down to Baltimore to meet her brother and get the keys to the brother's boat, which was docked up in Newport, Rhode Island, but on the way detoured to visit someone in Boston. On the fifth day my patient was just beginning to tire out, when they met a cousin of his new girlfriend. The cousin took one look at her and asked my patient: "How long has Jane been acting like this?" He said, "Acting like what?" To which the cousin replied: "Not sleeping, talking all the time, making plans to sail to Georgia, that kind of thing." My patient said: "She's a little high, but so am I; we're in love." At this point the cousin said, "Jane, when did you stop your lithium?" Turns out Jane was a manic-depressive and had not taken her lithium for two weeks. "*You* may be in love," the cousin said to my patient, "but I think she's manic again."

If love can look like mania, or at least early mania, I had wondered, could the biochemical processes be the same? This was as far as my speculations went, however, until another patient of mine, Joanne K., told me her story.

Joanne was a twenty-four-year-old writer whose mood states tended to alternate between very high and very low. Unfortunately, while the depressions tended to last as long as a month, the high (hypomanic) periods never lasted more than a week. Except when Joanne met someone during that week who really attracted her. "I frequently fall in love when I'm high" (she tends to look up old boyfriends at these times), Joanne told me. "Then I may stay high for as long as three or four weeks." Falling in love extended the length of Joanne's hypomanic periods, by providing extra juice for her synapses. Presumably it was the same stuff she was already burning, which meant the naturally produced chemicals norepinephrine or dopamine.

Then another thought struck me. When we treat depressed patients with antidepressant medications, a certain percentage of them tend to get a little too high and become hypomanic. This is not as serious as bringing on full-blown mania, which happens much less frequently and usually in patients with pre-existing histories of manic episodes. But what sometimes does happen during drug-induced, or even naturally occurring hypomania, which does get a little sticky, is that patients begin to get romantically involved with the wrong people. Like a young married schoolteacher's crush on one of her adolescent male pupils, or a forty-year-old father's passion for one of his teenage son's girlfriends. And with the drug-induced state we know that certain brain neurotransmitters or receptors are involved.

The prevailing theory about the biochemistry of depression involves downward alterations in certain neurotransmitter levels or activity. If our feelings of attraction are in fact tied to powerful surges in brain chemical systems, then at least for some types of depression, meeting someone you are attracted to should be a very effective antidepressant.

In fact, this turns out to be the case. While true melancholics cannot be budged by anything, falling in love is as reliable as any antidepressant for lifting many other depressives out of a slump. In fact, if they happen to meet someone while receiving medication, the two things may combine to make them too high.

There is one subtype of depressive that my colleague Donald F. Klein has labeled hysteroid dysphoric, who seem continually to fall in and out of live. When in love they are giddy, energetic, optimistic, and totally unable to view their partners realistically. They also eat and sleep less and are very sociable; the whole state is like an amphetamine high. When the romance ends, however, they crash into real slumps; at these times they overeat, oversleep, feel very sluggish, and don't want to see anyone. Interestingly, these symptoms very much parallel what people look like who are withdrawing from amphetamine. Whether you accept or reject the many psychological hypotheses about this or not, it is clear that meeting someone we find attractive who likes us as well gives us a lift. Which is why so many of us pursue romance so doggedly.

How the brain system fluctuations that accompany feeling attracted and attractive come about are still unclear, but it could involve some sort of amphetaminelike chemical whose level in our brains goes up when we meet the right person. As to what that chemical might be, we're simply not sure. Phenylethylamine (PEA) might be involved. As we discussed in Chapter 2, PEA is an amphetaminelike substance that may also be the source for norepinephrine and dopamine because it is chemically very similar to them. But could this substance itself play a role in our day-to-day emotional regulation, either as a naturally occurring antidepressant, or as a stimulant for romantic or other kinds of excitement?

This PEA theory of love has received a lot of publicity in the past few years, some of it very humorous. In one interview I remarked that chocolate was loaded with PEA, so perhaps people ate chocolate to enhance romantic feelings. This became the focus for an article in *The New York Times,* which was then taken up by the wire services, then by magazine free-lancers, and evolved into the chocolate theory of love. The problem is that PEA's role in our emotional life is at this point only a speculation. What we do have is the observation that drugs that raise PEA levels are helpful for treating people who habitually become depressed after a romantic disappointment (more about this in Chapter 6). Because of their response to drugs, as well as the nature of their symptoms, which mirror amphetamine withdrawal, these habitual and sometimes quite severe post-romance depressions might involve PEA deficits.

Many people do seem to eat chocolate when depressed. Could this be an attempt to raise their PEA levels? The problem is that PEA present in food is normally quickly broken down by our bodies, so that it doesn't even reach the blood, let alone the brain.

To test the effect of ingesting PEA, researchers at the National Institute of Mental Health ate pounds of chocolate, and then measured the PEA levels in their urine for the next few days; the PEA levels didn't budge. But perhaps measuring PEA levels in the urine is a poor way to tell about what's happening in the brain. Or perhaps we should test the people who say they get a lift from chocolate, to see if their PEA levels go up when they eat the stuff. But chocolate also has lots of sugar and caffeine, so perhaps the chocolate high has nothing to do with PEA.

A sharp reporter once asked me: "Okay, Dr. Liebowitz, suppose PEA (or some chemical like it) is involved in love. Do we put out more [PEA] when we are in love, or are we in love when we put out more [PEA]?"

That question stopped me in my tracks. As I think about it

now, it seems to be that the first thing we do is see another person. Something about the way that person looks, acts, or behaves must fit with our notions of what is attractive. It's not that the brain's cortex tells the limbic system, "He (or she) is attractive; crank up your motor." Rather, it's that when certain criteria are met, a limbic switch is automatically thrown, and we get excited. The saying "Beauty is that which is a joy to behold" means just this. Something that causes us to light up is seen as beautiful. When someone causes us to feel emotionally aroused by his or her very presence, we label that feeling romantic or sexual attraction, depending on the feelings involved. So to answer the reporter's question, whenever our limbic pleasure centers go bonkers in the presence of another person, whether this is caused by PEA, some other internal amphetaminelike substance, or another process altogether, we are likely to feel sexually or romantically excited. *what triggers pleasure*

I recently had an experience with a patient that made this more clear. He was a man in his thirties who had had a long history of falling for the wrong types, then getting very hurt. Treatment with a certain type of antidepressant over several years had helped him become relatively free of depression. However, this antidepressant imposed certain dietary restrictions, so we decided to see how he would do on lithium and switched his medication. A month later he came in to see me and said: "You know, this lithium is very interesting. While I was on the other drug I still kept getting turned on by people who were no good for me, even though I could handle it much better than I used to. With lithium I just don't get turned on by these people anymore."

This made perfect sense. We know lithium blocks manic highs, and also blunts the stimulating effects of amphetamine and cocaine. If people who are falling head over heels in love all the time are also having excessive and uncontrollable surges of the same biochemical systems, then lithium should help them as well.

We have talked about attraction so far mostly in terms of finding someone else appealing. However, it also seems that we feel stimulated by someone else finding us attractive or in some other way approving of us. This is the basis of the "applause" response. That we are biologically (as well as culturally and psychologically) programmed to find it rewarding to get approval from others makes sense for us as a social species. But approval is worth more or less depending on who is doing the approving. The more we value or esteem the source, the more rewarding their admiration feels. When it comes from someone to whom we are ourselves highly attracted, it is a powerful emotional experience.

Sometimes social approval seems to become an addiction, producing what I call "attention junkies," people for whom applause and admiration from others have become the most important things in life. Many seem desperately to pursue careers as performing artists even if they lack the necessary talent (which is not to say all performing artists are "attention junkies"). Others are constantly preoccupied with their dress or personal appearance. For many of these people, romantic relationships are highly troubled or unstable, because of an insatiable need for approval, which is the flip side of their constant fear of rejection.

How long attraction can last is something they don't teach us in the movies. What happens in the movies is that after a long struggle fraught with peril, the villains are defeated and the hero and heroine sail off into the sunset together. What happens after that is left to your imagination. And for good reason, because that is when the hard part actually begins. Namely, keeping a relationship together in such a way that it continues to be exciting. But more about this in a bit.

There are several popular alternatives to sailing into the sunset. One is for the hero to give up the girl, as Bogart does in *Casablanca*. This allows him to idealize her forever, and never to have to work on the relationship. Almost as good is the story of the star-crossed lovers whose fate it is to remain apart, such as Cathy and Heathcliff in *Wuthering Heights*. Sometimes fate operates through parents, as with Romeo and Juliet, and they kill themselves in despair. More often they continue to struggle, coming together passionately for brief moments every few years, while one or both are married to someone else, to tug at our heartstrings as we try to arrange outcomes in the book or movie so that the two lovers can end up together. The only problem is that we don't really know how well they'd do if they could actually remain together. Nor can they, because there is no way to know short of trying it.

ROMANTIC ATTACHMENT: HOW TO KEEP THEIR ATTENTION ONCE YOU HAVE IT

While the stimulant aspects of love strongly motivate much of our romantic behavior, it does not appear that nature has left it to our craving for excitement to insure intimate relationships with others of our species. For almost all of us to some degree, and for some of us to a great degree, anxiety about being alone also provides a powerful push toward getting and staying involved.

Anxiety or distress about being alone or isolated from others to whom one has become attached is called separation anxiety. Within weeks after birth, newborns of many species begin to yelp or cry when left alone, and don't stop, short of total exhaustion, until reunited with their mother or peers (depending on with whom they have been raised). This yelping or crying, which animal behaviorists call distress vocalization, occurs even if all of the young animal's other needs, such as food and warmth, are met, suggesting that they are specifically upset about being separated from their usual companions.

Separation anxiety has been extensively studied in primates, and the conclusion reached is that infants of every species react to being separated from their mothers with immediate agitation and powerful protest. The infants become panicky and frantic; their crying increases by as much as a thousand times. Heart rate, body temperature, and the presence of stress hormones such as cortisol rise quickly to peak levels, and remain high even if the animal is placed with a "substitute mother." This separation response can last from a few hours to as long as a week. If separation persists, some, but not all, animals may then go into what looks like states of depression, where they become apathetic and listless.

Anxiety over separation seems to be a built-in, instinctual, mechanism that serves the function of keeping helpless youngsters from wandering off. The more highly developed the species, the later separation anxiety seems to appear. Baby rhesus monkeys are said to protest their mother's absence beginning about two weeks after birth, while this is not seen in chimpanzees until three months of age. While there is controversy about when it appears in human infants, some studies have recorded crying when mother leaves and attempts to follow her in infants as young as four to six months of age; certainly by nine months many babies will show distress when their mothers leave and signs of pleasure, such as a smile, lifting of their arms, or delighted gurgles, when they return. Distress about being left by a particular figure, distinct signs of feeling comforted when that person returns, and resistance to the efforts of others to be comforting are all indications that a youngster has formed specific attachment to one person. Actually, they are pretty good markers for attachment in adults as well.

Separation anxiety is prominent in human youngsters for the first several years of life, but gradually curiosity and a wish to explore their world encourages children to tolerate absences from their mothers or other primary attachment figures. Yet residues of early attachment behavior and separation anxiety seem to appear in many forms in later life. We see them in our relationships with our parents as well as bonds formed with our peers, our children, and our romantic partners.

Some adults seem to react to separations as though they were still small, helpless children. It's normal to become attached to the person you live with, and to miss them if you're forced to be apart for any length of time. But some of the patients I've seen developed such attachment to their partners that it really became a problem. One woman threatened to kill herself if her husband, with whom she got along terribly, left her, mostly because she couldn't stand to be alone. Another, whose husband did leave her, couldn't stop calling him up even though he was abusive when she did. One man I saw would have an anxiety attack if his girlfriend didn't come right home after work; another couldn't stand for his wife to leave for five minutes. Clearly jealousy and possessiveness are present in these situations, but those feelings may be part of our attachment mechanisms.

If we are biologically programmed or wired from infancy to find social isolation distressful and contact with familiar members of our species comforting or rewarding, what might be the neurochemical basis of this set of responses? Specifically, what goes on in the brains of young humans or animals that causes them to cry when isolated or feel soothed when in the company of familiar figures? And could this childhood mechanism have anything to do with normal responses to social contact or isolation in adults? Could it, in fact, help explain overly strong or highly dependent attachments that some human adults seem to form? Recent studies and observations of both animals and humans are beginning to provide answers to these questions.

THE BIOCHEMISTRY OF ATTACHMENT IN ANIMALS

Anxiety over separation and the relief we feel when we rejoin familiar companions appear to rely on different biochemical systems than are involved in feelings of attraction. As we have said, the excitement of romantic or sexual attraction seems to involve amphetamine-related brain chemical systems. Attachment and separation anxiety, on the other hand, appear to involve the brain's own narcotics (the endorphins), and an area in the brain stem that seems to serve as a trigger for feelings of panic (the locus ceruleus).

First the animal data, for which I am relying primarily on work carried out by Professor Jack Panksepp and his colleagues in the psychology department at Bowling Green State University in Ohio. This research group is unique as far as I can tell in having studied the effects of a broad range of drugs on animal separation anxiety. Their strategy was to see if any type of drug specifically affected how animals reacted to separation.

What Panksepp and his colleagues found was that of many types of drugs tested in several animal species, only two had a specific effect on separation anxiety. One was a narcotic, which was able to greatly reduce crying or yelping (without making the animals drowsy) in puppies, young guinea pigs, and baby chicks who had been separated from their usual companions. A variety of other drugs, including stimulants, sedatives, antipsychotics, antidepressants, and tranquillizers could not do this without making the animals sleepy. The other drug that was effective in reducing separation anxiety was, interestingly, a medication used to treat high blood pressure, which specifically inhibits discharge of a certain brain areas. While not true in this work, other researchers have found that certain antidepressants reduce protests over being separated in young dogs and monkeys.

But how do you show that the brain's own narcotics, the endorphins, are involved? One way is to look at the effects of a narcotic blocker, which reverses the effects of both ingested narcotics and internal endorphins (and is used to treat heroin overdose). If endorphins do play a role in calming young animals when they feel isolated, or in soothing them when in contact with familiar figures, then giving them a narcotic blocker should have the reverse effects in both these areas. That is, it should make animals cry more when isolated, and feel less soothed when having social contact.

When young guinea pigs or chicks were isolated in pairs, where they still reacted to separation from their littermates but derived some comfort from the presence of at least one fellow chick or guinea pig, the blocker increased the amount of crying, implicating the brain's endorphin system in guinea pig and chick separation anxiety.

The next step was to test the effects of some manufactured endorphin, which turned out to be even more effective than morphine in reducing separation-induced crying.

What all this suggests is that animals (and perhaps also humans) are genetically programmed to secrete endorphins in situations of social comfort, which would then help them to feel both less anxious and also give them a sense of well-being. When they feel socially isolated or cut off from familiar figures, on the other hand, their brains shut off endorphin production and they go into a sort of narcotic withdrawal, which makes them feel anxious and even panicky.

HUMAN SEPARATION ANXIETY

Most adults feel some anxiety when separated from important figures in their lives, and some sense of increased security when their closest relationships seem stable. For example, you have a big fight with your girlfriend, and you may then find yourself worried that she will break off with you. So you call to apologize for your part in the argument, and feel a flood of warm relief followed by a sense of light-headedness when she says it was partly her fault and adds that she's looking forward to seeing you.

Separation anxiety may play a role in many aspects of romantic relationships. Some of the discomfort we feel when we are unattached may result from this mechanism; increasingly, many single people are turning to friendships to give them a feeling of connectedness. With time, some of the initial romantic excitement tends to fade from many relationships, which may then be held together by the growing sense of security that two people provide for each other. Some people form highly dependent attachments in which one or both partners literally fear to be alone. And some people make huge compromises, remaining in otherwise very unsatisfying relationships because of their fears about separating and having to be on their own.

To understand separation anxiety in adults, its helpful to look at what goes on in children. One of the world's leading authorities on early attachment and separation responses in children is a British child psychiatrist named John Bowlby. Starting out as a psychoanalyst, Bowlby has spent the last three decades observing and analyzing the effects of early life separations and reunions on children who for one reason or another had to be separated from their parents. His outstanding contribution has been to help legitimize the notion that humans as well as animals may be born with certain instinctual patterns of social behavior.

As described extensively in three volumes entitled *Attachment, Separation and Loss,* Dr. Bowlby has observed that a human youngster separated from his mother between the ages of one and three years characteristically goes through a three-stage reaction: protest, despair, and detachment. The protest phase is similar to that described above for young animals — crying to attract the mother, struggling to rejoin her, refusing to be comforted by strangers. In the despair phase the youngster becomes apathetic and listless; when he reaches the detachment stage he begins to function more normally but shows little response to human contact. In this phase separated children may not liven up for a time even if reunited with their mothers.

It makes sense that attachment and separation reactions have become genetically wired into us because of the aid to survival they offer — i.e., they keep young children close to home where they will be protected and cared for. Once a child can move about independently, however, his wish to be near his mother comes into conflict with his wish to explore his environment. More and more, a two-year-old will begin to take short trips on his own, frequently looking back and occasionally returning to make sure mother is still physically and emotionally there. Also, when alarmed, frightened, hungry, tired, or ill, the child's attachment will increase and his need for comforting will be greater. The more attentively mother watches, the more affectionately she greets him on his returns, the freer the child feels to explore. Should the mother begin to walk off, however, or rebuff him when he comes to her, the youngster will cling to her more and more and his explorations fall off.

Why do some children seem to show more separation anxiety than do others? In general, the more a mother (or mothering figure) does to provide the child with a secure base from which to venture forth, the more comfortable the child will be in doing so. On the other hand, the more the child has to worry about where mother may go when he leaves or how she will greet him when he returns, the less adventurous the child will feel. Bowlby believes that many children become clinging and timid because in one way or another they feel insecure about their attachments to their mothers.

The opposite extreme, overparenting, occurs when mother supplies too much initiative in approaching or comforting the child, rather than simply being available and responsive to his needs and cues. Such mothers fear to let their children go off on their own, constantly try to occupy their attention, and warn them of so many dangers that the children become frightened of the environment.

A leading child psychoanalyst, Margaret Mahler, has theorized that some adults fail to become independent because they have not been able satisfactorily to separate and individuate as young children. The principal reason that this may happen, according to Mahler, is that their mothers were unable to let the children comfortably separate from them, and did not indicate approval as they set off on their own or greet them affectionately when they returned. The effect of such a failure is to leave children plagued with one of two fears. The child may come to believe that becoming more adventurous or independent means giving up mother's love, which to a small child is terrifying. The wish to become more independent is therefore abandoned; one must remain closely tied to mother always to obtain her love. In later years this dependency will be transferred to others.

The second pathological possibility is that one gives up looking to mother for approval or returning to her for affection or comfort. Instead, she comes to be seen as someone who wants to smother the child, who does not want to surrender his drive for autonomy.

In later life all intimate relationships are seen as potentially engulfing and avoided like the plague.

Mothers seem to receive most of the blame if their kids turn out to be clinging or distrustful. But what must also be considered is that differences in how independent or dependent we are could be inherited. Some of us may just be "clingier" than others. The same maternal encouragement may send one child off and running while another still feels insecure and frightened to explore. The problem is that how mothers relate to their children has been easier to study than how children differ in their innate or inherited temperaments.

The biological basis of human separation anxiety has received very little study. But what has received some attention are adult panic attacks. Some people seem vulnerable to panic attacks, which are brief episodes of intense anxiety that are not provoked by a realistic threat. In their emotional and physical symptoms, panic attacks are similar to child separation responses. While in some cases panic attacks appear to begin after a severe medical illness or a drastic hormonal change, for others the trigger is the loss of an important relationship. Also linking panic attacks to separation anxiety is the fact that twenty to fifty percent of adults who develop panic attacks had intense separation anxiety as youngsters, as evidenced by trouble going to school (really a fear of leaving home), going to sleep-away camp, or staying at a friend's house overnight. Further supporting this link is the fact that the same medications which help adult panic attacks are also effective in treating school phobias in children, which most often really involve the child's fear of being separated from a parent rather than fear of school per se. Adults with panic attacks often can't go out alone because of their fear of becoming panicky, but can travel in the company of a trusted companion, suggesting that the biological effects of the social contact help reduce vulnerability to panic. People with panic attacks also often become intensely dependent on others, usually a spouse, and constantly worry that they will be abandoned. If the panic attacks are treated with medication, however, often this pattern of dependency changes and the patients become much more independent. Sometimes after treatment they end up leaving the spouse they formerly thought they couldn't live without.

All this is illustrated very nicely by the story of Mrs. J., a woman in her early forties who was married to a man she was no longer romantically or sexually interested in. Because of a long history of panic attacks she felt unable to work and therefore unable to leave her husband, on whom she depended financially. Soon after starting antipanic drug therapy, her anxiety attacks ceased. After several months she began to look for a job, soon found one, and separated from her husband.

It is also interesting to note that there is a strong family predisposition to panic attacks; the close relatives of a person with this disorder are many more times as likely to have the same problem as are people in general. While a family pattern for an illness doesn't prove it's genetic, it is food for thought.

People who get panic attacks also seem to have faulty alarm circuits. If you give a heavy intravenous dose of a certain salt solution (sodium lactate) to most people, the worst they experience is numbness, tingling, and perhaps some twitching. But give it to someone who has panic attacks, and he or she is very likely to hit the roof with panic. Whatever the original cause of their anxiety, something about their circuitry has become abnormal, and alarms go off when they shouldn't.

While we don't know for sure, we think that a specific part of the brain (called the locus ceruleus) operates as an alarm center in human beings and discharges unpredictably in people who get panic attacks. This alarm center can be switched off by narcotics, by one particular type of high blood pressure medication, and by certain antidepressants (even in people who are not depressed). It would be extremely interesting to give adults with recurrent panic attacks a narcotic like morphine to see if it helped, but this is not ethical because it involves the risk of addiction. But our own research has shown that other, non-addicting drugs that switch off this alarm center are helpful in treating panic attacks. We have also shown that if someone who gets a panic attack from sodium lactate is then treated with medication, the lactate no longer can make them feel panicky.

To sum up, medications that inhibit our brains alarm center reduce protest over separation in young animals and block panic attacks in humans. It would seem, therefore, that one can make a good case that this same alarm center is involved in both animal separation anxiety and human panic attacks. If so, it would suggest that we are biologically wired so that certain brain anxiety circuits become more active in situations of social isolation and quiet down in situations of social comfort. Whether the mechanism involves our internal narcotics, the endorphins, as seems to be the case in several animal species, is not yet clear.

We do know that narcotics, and presumably endorphins as well, produce both calming and euphoric feelings in humans, and that even regular narcotics users continue to experience at least a brief euphoria with each drug administration. Thus the effect doesn't wear off, which would have to be the case for something that helps us remain interested in long-term relationships.

One could also hypothesize that some people get addicted to relationships just the way they do to narcotics, and that perhaps these two processes were in some way linked. The interaction between the mechanisms of pharmacology and our intimate relationships is more extensively discussed in the next chapter.

Separation anxiety is a normal part of human romance, giving rise as it does both to our anxieties when our partner is away or the relationship seems troubled, and the sense of security we feel when we are together and things are going well. However, for reasons that may be biological (e.g., the instability of their alarm system circuitry) or environmental (e.g., a mother who stifled any early attempt to explore or become independent), some people seem to have more difficulty with separation anxiety in their romantic relationships than do others.

In the extreme such people do anything to avoid being alone. As a consequence they often tend to be less choosy about whom they get involved with, as long as the partner will not make too many demands on them to be independent. These relationships aim for security, not excitement. The faster they get married, buy the house, and have the kids, the better.

Even with a very reassuring partner, this fearful lover wants constantly to be told that he or she is still loved, which is re-assurance against abandonment. To want such assurance occasionally is pretty common; it is the constant need and the inability to believe whatever is offered that is excessive. Such relationships tend to be filled with jealousy and possessiveness. One highly separation-anxious patient whom I saw in consultation had actually gotten to the point of being afraid to let his wife out of his sight. Even when she went downstairs to the mailbox alone he imagined that she was having an affair.

People who desperately fear being alone sometimes have to put up with a lot of abuse from their partners. If you can't tell someone to take a walk when he or she mistreats you, you have lost a lot of bargaining power in your relationship. It's nice to imagine that lovers always have their partner's welfare upper-most in mind, but that is a fairy tale. Successful love relationships at times require that you negotiate to get some of your needs met, especially when they may conflict with your partner's needs. The most extreme thing you can withold is your presence, by refusing to stay around. In certain situations, your partner must know that you will put up with only so much and then it's goodbye. But it's no good to threaten it if you don't mean it or are unable to carry through. Which means that you have to be able to make it on your own before you can really hold your own in a relationship.

A SPECIAL ROMANTIC STATE

There are certain aspects of a romantic interaction that at times begin to take on some of the qualities of a mystical experience. While most people don't usually think of their most profound romantic memories in these terms, from a neurochemical perspective I believe the term is appropriate.

What I have in mind are those moments when a person begins to break through his or her own normal ego boundaries and to experience a sense of or desire for merger or oneness with a partner. Newly formed couples who gaze into each other's eyes for long periods of time may experience this. It is also a feeling that may occur during sexual intercourse, or when you have just shared something very personal. These feelings seem to occur at the times when our love or passion is most intense. At such moments we may be torn between a desire to stay with the experience and a wish to back off and regain a little distance.

These are the peak moments of our relationships, the times we never forget and yearn to recapture or experience again. While never studied biologically, they have many features in common with overwhelming aesthetic or powerful religious or mystical experiences. Human beings seem neurochemically wired to hit such emotional peaks occasionally, and moments of great intimacy with another human being can sometimes flip the right switches in us.

These peak love experiences are both intensely exciting and calming, but also something more. The difference between such a moment and more ordinary but still powerful romantic experiences is like the difference between taking a psychedelic drug and a more ordinary stimulant or narcotic. This leads me to think that these peak love experiences involve such intense stimulation and joy that some additional neurochemical reaction is triggered, which may be similar to (although usually briefer than) whatever the psychedelic drugs do to our brains. What then happens is that our normal ways of experiencing ourselves and the world around us begin to change. This can involve a temporary loss of the sense of your own boundaries. It can also involve an experience of oneness with your partner or a sense of timelessness as though your feelings will last forever. At such moments, one experiences a wonderful sense of completeness and a wish to remain in this state forever.

There are two ways of looking at this. On one hand, perhaps many of us have some deep-seated yearning to merge with someone or something, to share ourselves totally with another person, to break down the barriers that separate us from other people. When we get close to that, as may happen in certain very romantic moments, we feel incredibly excited. However, this may make some people uneasy as well, because at such times we are stripped of our usual defenses and are therefore very vulnerable.

The alternate explanation is that in moments of great excitement or joy, whether produced by a painting, a sunset, or a romance, our brains go a little haywire and we begin to have unreal experiences. One of the resulting short circuits disturbs the normal processes that govern our sense of boundaries, so we may feel like we are merging with time, the universe, or the person lying next to us. On these occasions the experience is pleasant, not because we inherently want to merge, but because of the original event that produced our excited state in the first place. In fact for many people this sense of merger and loss of ego boundaries is scary, and they may avoid intense relationships or interactions to keep from having such experiences. Also, if you just happened to be standing around and felt yourself becoming one with the room, you'd probably run like hell.

Whether we actually want the sense of merging, or just the excitement that sometimes produces it, is hard to figure out. What does seem clear, however, is that people differ in what they find exciting and therefore in the kinds of relationships they pursue. For example, some emphasize love as the basis for relating to the opposite sex, and disparage the casual liaisons of the swinger set. Others feel just the opposite, and believe that variety, not commitment, is the spice of life. Are these two types getting the same emotional buzz from their different ways of relating? Or are they seeking different types of excitement, the lovers balancing attraction and attachment with an occasional peak romantic moment, while the swingers, into attraction but avoiding attachment, prefer the wild times that are most like, as well as often mixed with, the effects of alcohol?

People differ markedly in how easily they fall in or out of love, in how intense a romance they prefer, and in how strong or long-lasting a commitment they tend to be comfortable in making. While any particular person's overall approach to romance may vary somewhat depending on who he or she is involved with at the time, there are still certain constants in how any one of us has and will relate to the different partners in our life. These constants, which make up our own particular pattern of loving, are strongly shaped by our individual attraction and attachment styles.

DIFFERENT TYPES OF LOVERS

In an article entitled "The Styles of Loving," John Alan Lee described six distinct patterns of feeling and behavior that his research subjects reported in their love relationships. Lee labeled these patterns Eros, Ludus, Storge, Mania, Pragma, and Agape.

Erotic lovers get taken with each other very quickly, go to bed soon after meeting, and demand physical perfection. Erotic lovers are principally concerned with beauty; personal and intellectual qualities come second. Once a suitable partner is found, there is a strong desire for intimacy. However, erotic lovers are not needy, possessive, or afraid to be alone; rather, they tend to be self-assured types who are willing to pursue an ideal.

Ludic lovers are playful types for whom love is a pastime. They don't get deeply involved emotionally, and don't want their partners too involved with them either. Often they date several people at the same time, and take care not to see anyone too often.

In contrast to the intensity of Eros and the playfulness of Ludus, storgic love develops slowly, quietly, and without great passion. It is more friendly or even siblinglike, and tends to grow between people who live or work near each other, between whom an increasing fondness grows over time. Going to bed occurs late in the relationship, and marriage, home, and children are usually the goals.

More like erotic lovers, but less stable and independent, are the fourth type — manic lovers. Manic lovers get consumed by thoughts of the other, experience incredible highs when things go well, and sink to the depth of despair if the partner seems not to be responding fully. Manic love is like an emotional roller coaster. It's the stuff of romantic literature, full of passion, jealousy, and possessiveness. Manic love is also what I see in a lot of my patients.

Next comes Pragma, which Lee calls "love with a shopping list." Pragmatic lovers choose each other to be compatible in their interests, background, and personalities. Intense feelings can develop with time, but the primary consideration is a practical match. Many arranged marriages are of this type.

Finally there comes Agape, which demands total compassion, altruism and nondemandingness in the lover. Lee calls it the classical Christian view of love, which may be why he found no pure examples of this type in his research sample.

Now let's look at these patterns in terms of what we've already discussed about attraction and attachment. When you meet someone and get a quick stimulant rush, you feel very attracted. This is what happens in erotic lovers and in manic lovers. Where they differ is that manic lovers also get very quickly attached, so they panic or crash if the other person starts to pull away. For erotic lovers, the attachment process comes much more slowly. Ludic lovers also seem to get at least a moderate stimulant rush, but don't get attached at all.

The other possibility is for there not to be much emotional (or chemical) attraction at the beginning but for the attachment process to develop anyway. This seems to happen with storgic lovers who start off as friends, and pragmatic lovers, chosen because they will work out well in the long run. Agape doesn't fit any of these patterns, which is probably why it doesn't exist.

We need a lot more research on how and why people differ in their attraction and attachment patterns, but I have a few ideas. As we've said, chemically, attraction is basically a stimulant system. People with low MAO seem to have, and to prefer, more intense emotional stimulation in their lives. People whose brains don't damp down intense stimulation (who are called AER augmenters) may also be like this. Romantic preferences also seem related to how much and what type of sensations we prefer — whether we are particularly thrill oriented, prefer wild parties, or value deeply felt emotional experiences. Perhaps people who tend to feel quickly and powerfully attracted romantically differ from those who don't in their MAO or AER, or in some related biological system. How we react to stimulant drugs might also tell us something.

Attachments are more security than excitement oriented, and how quickly or strongly we form them may also have to do with our biological wiring. Some of us may be more attachment prone, which could have to do with inherited differences or early life experience. This might show up in how much independence we sought as children. Someday we might also have biological tests to measure our attachment and alarm systems. These could involve testing the effects of a narcotic blocker (or a narcotic if a nonaddictive one is found) on our feelings of attachment, or how sensitive each of us is to drugs that stimulate brain alarm systems.

This may sound like science fiction now, but there may come a day when, if you wanted to learn more about yourself as a lover, you would have to get to know your biological self. The same would hold true if you wanted a better idea of what made another person in your life (or someone you were thinking of bringing into your life) tick romantically.

Abraham Maslow's ideas about love, although again exclusively psychological, also lend themselves nicely to our discussion. Maslow identified two contrasting types of love, which he called D-love (deficiency love, needing love, selfish love) and B-love (love for the being of another person, unneeding love, unselfish love). D-love, related to normal infant needs, is experienced by adults as a hole to be filled, an emptiness for someone to pour love into, a hunger that can be only temporarily satisfied. The partner is valued for his or her ability to satisfy this hunger for love.

In contrast, B-love involves more of a desire to give than to receive love, is less possessive and more enjoyable, less needing and more admiring. On one hand B-lovers can do without, yet also do not become satiated. B-love is an end rather than a means, and is often described as being the same as, or having the same effects as an aesthetic or mystical experience. D-love involves more anxiety and hostility, although B-love can involve anxious concern for the other's welfare. According to Maslow, B-lovers are more independent of each other, more autonomous, less jealous and threatened, more generous and fostering of their partners.

Maslow recognized that most people experience both types of love at the same time but in varying combinations. While there are undoubtedly psychological factors determining which type of love predominates in a given person's emotional repertoire, a biological perspective helps explain what is going on in the two different love states. Pure B-lovers are people who for one reason or another do not experience any unmet attachment needs or anxiety about separation in their romantic relationships. This could be because they don't get readily attached to people in general, or because they are not very attached to the particular partner with whom they are in a B-love state, or because they feel so securely attached to their partner that there is simply nothing to worry about. In any event, they are free to indulge in the more playful aspects of romance, involving the stimulant aspects of our limbic neurochemistry.

D-lovers on the other hand are very concerned about their attachment, either because it is in fact tenuous or because they have such strong needs in this area that no one, no matter how giving, can make them feel very secure. In either case, they live in perpetual states of anxiety and concern over possible abandonment, driven by unrest in the alarm mechanism of their separation anxiety circuitry.

When one begins to link emotional relationships to changes in brain chemistry, an interesting question emerges. Do people seek out loving relationships in order to bring about changes in their brain chemistry, or do the changes in brain chemistry help facilitate people getting or staying involved with each other? Actually, I'm not sure that there is a clear answer to this question. From an evolutionary standpoint the answer is not important; the principal concern for our survival as a species is that we do seek out partners of the opposite sex, mate, and stay together long enough to insure survival of the offspring. But to romantics, the idea that their partners may in some ways be viewed as mood-adjusting agents must surely seem sacrilegious.

ARTIFICIAL SWEETENERS

Any time you fall passionately in love with someone a certain amount of fantasy is involved. Naturally some degree of disillusionment is inevitable as, with time, you get to know your new partner. But we all hope the initial appraisal wasn't too far off, so that the person you wind up with when the honeymoon is over bears some resemblance to the person you thought you were getting involved with in the first place.

There are certain situations, however, in which we are so keyed up emotionally that we can become powerfully attracted or attached to people with whom we would not normally get so involved, such as wartime (or even holiday) romances, meeting someone just after a painful romantic breakup or after you have been isolated from other people for a long time or been very ill. In situations like these, attraction and attachment systems become supercharged, and a partner who in normal circumstances would not engage you is suddenly very stimulating or comforting. The problem is that more than likely you are not going to spend the rest of your life in that emotional state. As your thresholds readjust, your new partner may not continue to seem so fascinating.

The crux of the matter is that love is not a one-time thing. No matter how intensely you felt about so-and-so at one point in your life, he or she must remain interesting or in some way emotionally satisfying for the relationship to stay viable.

This is why there is no such thing as a "love drug" or a love potion, unless you are going to stay chronically intoxicated all your life. Drugs like amphetamine or cocaine do lower pleasure center thresholds and may make someone seem more appealing initially. But when you come off the drug your biology reverts back to normal, and so will your romantic preferences. In short, the closer you are to your everyday self when you fall in love, the more chance you have of picking someone who will continue to interest you.

Sex may be confusing in this way as well, because it can be so arousing that intense sexual attraction can sometimes be confused with love. If it's love, then even though your sexual drive may be temporarily gone, other aspects of your attraction and attachment should still be operating. If it's pure lust, chances are you'll feel much less.

There are several other situations that heighten interest in or attraction to another person but don't really have much to do with that person per se. Overcoming an obstacle may heighten or even cause an attraction for some people. Someone who plays hard to get can present such a challenge. Someone already in a relationship, or even a marriage, can also present such a challenge. Someone your parents disapprove of, or whose parents disapprove of you, can present such a challenge. We'll look more closely at people who are so motivated in Chapter 5. Whatever the motivation, these obstacles can heighten attraction because they generate extra excitement as we struggle to win over or possess the other person. As such, they can lend a romantic aura that may not hold up when you finally have the partner all to yourself.

Experts in the field have for some time debated whether the feeling state we call passionate love has one or two components. The one-component theorists argue that love, or intense attraction anyway, is a distinct feeling state with its own pattern of physical arousal and mental reactions. If some amphetaminelike substance does get squirted in our brains when we see someone we like, and squirted more if they seem to like us as well, a one-component theorist would not expect to find this chemical sequence with other emotionally charged kinds of encounters, such as when we feel scared or threatened. Two-component believers, on the other hand, say that all intense arousal is physically the same, and what emotion you have depends on what label the thinking part of your brain assigns to it. If we feel aroused and see a pretty girl, it must be love, or at least lust. If instead we see a charging rhinoceros, we call it fear.

In their book *Interpersonal Attraction,* psychologists Ellen Berscheid and Elaine Hatfield Walster cite several experiments that bear on this issue. In one, a group of male volunteers were told that they would soon receive three "pretty stiff" electrical shocks. Half were later told this was not going to happen after all. A separate control group was involved in the experiment but never told anything about shocks. All three groups were then introduced to a young female college student, and then asked how much they liked her. The experimenters predicted that those still scared about the shocks and those who had been scared but then let off the hook would both be more physically aroused than the men who had heard nothing about shocks, and that their aroused state would make them think they liked the young woman more than would the men in the third (control) group. These expectations were apparently confirmed in the experiment.

In a second experiment, a group of college men were given some personality tests. They were then given false results; some were arbitrarily told they showed many positive traits, others that they showed very few. Soon after, each man was approached by a young woman who was part of the experiment. To half the men she acted warm and affectionate; to the other half cold and rejecting. The results: the men who had received unflattering personality evaluations were more attracted to the girl when she behaved warmly, and more offended when she behaved coldly, than were those who had been rated highly by the experimenters.

While this experiment does show that a painful experience influences how we respond to people, it does not support the idea that any emotional arousal, whether pleasant or painful, sets us up to feel more attracted. A more reasonable interpretation is that the men who were put down by the researcher were still licking their wounds, more touchy about how people treated them, and more appreciative of anyone who treated them nicely.

As part of a third experiment, a group of male volunteers were asked to fill our questionnaires for what they were told was a computerized dating service. Thereafter they were all given the phone number of the same woman, who was actually part of the research team. To some of the men she responded warmly and appeared grateful to be asked out; with the other men she acted less friendly and more reluctant to accept a date. The researchers predicted that going out with someone who was hard to get a date with would be more arousing than dating someone who was easy to get a date with, because it would seem like more of a challenge or accomplishment. Therefore the men would report themselves as feeling more attracted to or excited at the thought of seeing the "hard to get" date. The results did not bear this out, because the men reported themselves to have an equally high opinion of the woman no matter how she acted when they asked her out. Tying this together with other research, the bottom line seems to be that men do prefer women who are hard to attract more than women who are easy, but only if it's not hard for them. That is, men want someone who is hard for other men, but easy for them, to attract.

Someone who gives you a hard time is not as appealing as someone who is difficult for others but easy for you to get, because anxiety and fear per se are not turn-ons. It would make no sense evolutionarily if they were, because we would constantly be exposing ourselves to dangerous situations. We feel good when we get approval and acceptance from others, which is what makes us capable of living together as a social species. We care how other people think of us, and like to be admired. The more we value another person, the more we value his or her admiration.

Our emotional responses are composed of cortico-limbic programs to which we react reflexively, and then with secondary rational elaborations. There are basic circuits that are built into us from birth, and when something triggers them they go off. Where the assessment of cues may play a role is how we fine-tune one of these basic states. I'm very excited — is it only sexual or something more? I'm angry — should I only be annoyed or really let them have it? I'm scared — do I try to hold my feeling in check and carry on or do I let myself panic and run like hell? The cortico-limbic program itself is a product of our instinctive wiring that has been changed and modified over time by those experiences stored in our memory banks. Thus for each of us it will be different, which is something none of the experiments cited above takes into account. One person may laugh when the experimenters arbitrarily tell him his tests show he is immature, while for someone more vulnerable this can be crushing.

Beyond our programmed ways of responding, each of us uses our mind and the available cues to fine-tune our feelings. Depending on my current life circumstances, past experiences, and future hopes, I may label in a variety of ways the arousal I feel after an animated discussion with a woman I find attractive. If I am married and committed to fidelity, I may call it mostly intellectual excitement, or in some other way try to minimize or channel it. If I am single or recently divorced, I may do the opposite, and let my romantic or sexual fantasies embellish it, at least for a while. What it comes down to, finally, is that romantic feelings are not totally specific nor just nonspecific arousal which we may then label as one feeling or another. Rather, they involve our highly individualized feelings of attraction and attachment, which are biologically, psychologically, and culturally shaped, plus our secondary intellectual fine-tuning. While this may sound complex, we should be consoled. It tends to happen instantaneously and involuntarily — at least the first reaction. And it does make for variety.

CHAPTER 5

Romantic Difficulties

My principal point so far is that our romantic ups and downs have important biological aspects, and that the particular brain systems involved are the same as in our reactions to certain powerful drugs, other heady emotional experiences, and even some of the psychiatric disorders. If this is true, then a number of things follow. For one, the biological principles governing how drugs affect us should also apply to romance. Also, then at least some of the romantic troubles that so many people experience should make more sense when a biological perspective is added. Finally, drug therapy should help some types of romantic difficulties.

Let's look at each of these issues more closely.

HOW OUR BRAINS HANDLE ROMANCE

Earlier we discussed how drug research suggests that our brains react most strongly to substances to which we have not been previously or at least not recently exposed. If this were true for how we react to people as well as to chemicals, it could explain why new partners can feel so stimulating, at least for a time, and why our fancies are so often caught by passing strangers. Most people involved in long-term relationships with one person are attracted to someone else from time to time, whether or not they choose to pursue it. The newness can make these strangers seem more glamorous than our regular partner who is more familiar to us, but how that glamour would hold up if we actually pursued our fantasies is another matter.

The biological mechanism for this newness or novelty effect most likely involves the power of an attractive but unfamiliar person to elicit a huge surge in our brains, which leaves us feeling very aroused. Whether this amphetaminelike rush allows us powerfully to idealize complete strangers, or whether we can begin to idealize them because we don't know them, and then get turned on by our fantasies, is hard to say. In either case we end up seeing them through biochemically colored glasses.

The power of unfamiliarity to enhance sexual attractiveness may have been advantageous in evolutionary terms, since those in whom this tendency was strongest would have been sexually active with more people, ensuring a greater chance of survival for their genes. This could explain how it became widespread in our species, as well as in many others. There is a joke about President and Mrs. Calvin Coolidge touring a chicken farm. They see a rooster running around a coop mounting one hen after another. Mrs. Coolidge then says: "Mr. Coolidge — do you see how active that rooster is?" To which "silent Cal" retorts: "Yes, Mrs. Coolidge, and do you see how many hens he is active with?"

Many enduring relationships are based on novelty, which may at first sound contradictory. The only requirement in such cases is that the couple not try to live together; in fact, the less prolonged time spent together, the better. Many such intermittent romances can go on for years. People who live apart and see each other occasionally, people who live with one person and clandestinely see another, or people who come together and break up repeatedly are all examples where the attraction is fueled by novelty, and might not survive a more steady exposure.

No one has compared the love life of sensation seekers to that of non-sensation seekers, but one would predict that the former might have more trouble sticking with a single relationship because of their craving for the intensity that comes with novelty. We would therefore see greater reluctance to accept the structure that marriage imposes, and a greater tendency toward extramarital affairs or divorce in those sensation seekers who do marry. As we discussed previously, people with low MAO enzyme levels seem more prone to certain extreme kinds of behavior. It would be interesting to see if they were also more prone to marital instability.

Related to the issue of novelty is the "ramp effect." When we talk about ramp effects in pharmacology, we mean changes in the level of a drug in the body. The steeper the ramp — that is, the more rapid or greater the change in drug level — the more powerful its effect on us. Once the drug hits its peak concentration and levels off, its effects begin to subside, even if the amount still present in the brain remains high for some time. Exposure to a constant amount has less and less effect; change is what we respond to.

This is one of the principles behind the honeymoon, although it is getting less and less true as an increasing number of couples live together before marriage. Traditionally a honeymoon was the first opportunity a couple had to spend a prolonged amount of time together; in chemical terms you could say they got a big dose of each other all at once. Busy working couples treasure weekends or vacations together for the same reason; it is a big change from the small amounts of time they usually have with each other. Fighting also helps, especially if you then kiss and make up, because this gives you a nice big change in your intimacy level all at once.

Try spending that same amount of time together on a regular basis, as happens when people retire, and the glow may quickly fade; people who used to want more time together all of a sudden look for excuses to go out alone. The reason is there is no longer any change or novelty. Instead there is a steady exposure to the same level of contact. While this does not need to be a difficulty, it usually takes some work to keep things interesting.

Underlying the power of both novelty and sudden change is the mechanism of tolerance. Becoming tolerant to a drug means becoming used to its effect, so that the same dose has less and less effect over time.

Tolerance is a real problem in romantic relationships, because it leads to people getting bored with each other. The problem becomes worse the longer a relationship lasts, unless one or both of two things happen. The first is that with time we may become more attached and comfortable with each other, and willing to sacrifice excitement for security. The second is that we find ways to put some novelty and change into our love life. Fights do this, because we can make up after they are over. Vacations do it. Unfortunately, at least from the standpoint of marital stability, affairs and divorces also do it. This is really an important issue and we will discusss it more thoroughly a bit farther on.

Novelty, ramp effects, and tolerance are all involved in what I call the "old boyfriend" or "old girlfriend" syndrome. Frank T. was a case in point. Good-looking, successful, charming, and in his early thirties, he'd gone out with Beverly for several years, but eventually got tired of the relationship. Soon after, he met Jill. They dated for a while, fell in love, became engaged and then married. At first Frank seemed extremely happy, and commented on the things that he and Jill could share that had been missing from his previous relationship. After a while, however, some of the excitement seemed to fade, and he found himself fantasizing about seeing Beverly again. It was not hard to find an excuse to call her, nor to find one to get together. Soon they began having an affair. At this point he said being with Beverly was terrific.

What goes on in this and other cases like it is that with time and constant exposure interest fades, whereas after a long interruption an ex-lover can seem very new again. Add to this the intermittent (and clandestine) nature of the affair, and the revived romance can seem very glamorous . . . as long as the old lovers don't try to get back together on a permanent basis, because it usually doesn't work. Frank didn't try, so I don't know if it would have worked out for him.

Where you take a drug, or a date, is important. A setting is something that establishes a mood. Lovers have known this for years, with walks along the shore or through a park, or dinners over candlelight. We know that settings alter brain thresholds for drug effects, and chances are they do the same for romance. Nostalgic settings rekindle passions in old lovers, just as former addicts begin to experience drug cravings when they visit their old haunts.

Cravings are related to the phenomenon of withdrawal. Withdrawal is what you feel with certain drugs if you've taken them for a while and stop suddenly. Withdrawn is what you often feel if a romance has just come to an end. Are the two linked? Success with a lover increases the hyperactivity of certain neurochemical systems, which enhance feelings of excitement or security. When a relationship ends, the stimulus for those brain effects is lost. Reminiscent of the ramp effect, but perhaps more appropriately called the slide effect, the more sudden the loss of the person, the more sudden the loss to the brain. What we slide into is some balance of sadness, which is most likely the sudden wearing off of a high brain stimulant state, and anxiety, which could come from the sudden loss of chemical (endorphin?) inhibitors of our anxiety circuits.

Some readers may say that, after all, when a relationship ends what we have lost is the presence of the other person. That is quite different from having a chemical leave our bodies.

Actually, however, when a person leaves, what most changes is the mental image of ourselves in a relationship with that person. This is what gives rise to the sense of loss. If someone leaves in the middle of the night and we don't know it, we don't feel abandoned. If we are separated from a lover for a year but still feel that the bond is strong, there may be loneliness, but no sense of loss or mourning for the relationship. The point is that in any relationship we spend much more time alone but feeling bonded than actually interacting with the other person. And when a relationship ends, we spend a lot of time alone thinking of ourselves no longer being able to interact with that person. One mental script is substituted for another.

But when does the chemistry come in, you may ask? The answer is that these scripts have their emotional impact on us by powerfully affecting our brain's limbic system. It appears that we are wired from birth to feel joy when we feel strongly bonded, and to feel agony when these bonds are disrupted. But the joy or the agony is due to neurochemical changes in the brain's emotional circuits induced by what our brain tells us is going on at the moment. Thus life events, just like drugs, have their emotional impact on us via change in brain chemistry.

How long we feel terrible after a romantic loss is a complicated matter, and depends on the resiliency of our neurochemistry as well as our social supports, outside interests, and prospects for meeting new people.

With time, most people recover from a broken romance. Some actually bounce back a little too vigorously, and wind up impulsively marrying someone whom under normal circumstances they might not find overwhelmingly attractive. These folks are said to be marrying on the rebound.

When we talk about rebound effects in pharmacology, what we are referring to is the neurochemical backlash that sometimes occurs following prolonged use and then discontinuation of certain drugs. What is thought to happen is that receptor sensitivities change. If constantly exposed to stimulants, certain brain chemical receptors become less sensitive, so that right after the stimulants are stopped they, and we, are relatively unresponsive and sluggish. With prolonged exposure to sedatives, the opposite process occurs, and certain activating brain receptors become supersensitive. As a consequence, the postdrug state is one of hyperexcitability, which can lead to restlessness, jumpiness, and even convulsions.

The analogous process in romance is when the brain's attraction or attachment drives turn off for a while immediately after a relationship, but bounce back with a vengeance after an extended period of social withdrawal or isolation. If our biological wiring is similar to that of other species, then when we are suddenly deprived of an important social bond our brain endorphin activity temporarily shuts down. With the passage of time, however, the system begins to prime itself again as our receptors become more sensitive and storage pools increase their stocks. Along comes a somewhat attractive and friendly soul and whammo, our brains are hit with megadoses of "attachment juice."

These theories are offered with the proviso that they be regarded as speculative, in need of verification or disproval by experimental testing. So anyone whose romance is on the rocks should not start reaching for the endorphins unless perhaps he or she is also willing to take a matching placebo so we could turn it into a research study. But it does make sense to see your friends as much as possible at these times to fulfill some of your unmet attachment needs, and perhaps prevent whatever brain system is involved from going into complete shutdown. It also makes sense to realize that you can become especially vulnerable as you begin to mend and therefore need to look hard before you leap too far into a new relationship.

Drug dependence is a situation in which the body goes through a state of physical withdrawal if use of that drug is suddenly stopped. Many animals can be made dependent on a drug, but will not seek to use that drug again once it is out of their system and they are through with the withdrawal effects. This is the difference between dependence and addiction. Addicts tend to get hooked again and again.

Although it may seem strange to think about it in these terms, many if not most human beings seem to become neurochemically dependent on their important relationships, so that they go through withdrawal symptoms if these ties become unstable or suddenly end. This can take the form of the rapid heartbeat and sweating that accompany anxiety, or the fatigue, lethargy, or agitation of depression. We tend to think of these as secondary to the emotions we experience in such situations, but then the emotions themselves are linked to chemical changes in our brains related in turn to the interpersonal threat or loss. One might therefore define dependence in our personal relationships to include any which put us through intense emotional changes when they end.

In effect I am challenging the distinction that pharmacologists and lay people alike make between physical and psychological dependence. Physical dependence means you suffer strong physical symptoms if you suddenly stop a drug; these tend to wear off fairly quickly. Psychological dependence means that you don't suffer the same physical symptoms but still feel crummy without the drug and work like heck to get some more; also, these symptoms can last much longer. Marijuana is an example of a drug upon which some chronic users are said to develop more psychological than physical dependence. But if we think of marijuana's euphoric, anxiety-relieving, and antidepressant effects in neurochemical terms, then the restlessness, anxiety, and depression a chronic user may feel it he runs out of the stuff are really due to chemical changes going on in his brain.

Looked at in these terms, any relationship we will not surrender willingly is one on which we are dependent. Sometimes this is highly adaptive; for example, the efforts a parent will make to aid his or her child in distress are fundamental to the child's survival. In romantic relationships, the same measures might apply — how dependent we are can be assessed from how hard we struggle to maintain the relationship in the face of difficulty, as well as what we suffer if it ends.

As we've said, dependence is not the same as addiction. Addicts seek to use drugs in ways that bring a sharp pulse of the substance to the brain. They also become dependent on the stuff over and over and over again. Using these two criteria, can we say that some people are addicted to romance?

The evidence suggests they are, although an attraction addict may look different from an attachment junkie. Attraction addicts relentlessly pursue the experience of falling in love, or continually want people to fall in love with them, or both. Attachments don't form, or don't last very long if they do, unless it's to someone who can give the attraction addict a great deal of freedom. Otherwise, being attached is an obstacle to chasing new attractions. Attachment addicts, on the other hand, dive quickly into highly dependent relationships; if one doesn't work, they find another. What they always seek is to find someone they can cling to and with whom they feel secure. Such people are not looking for excitement, but are driven by a desperate need to escape anxiety over being alone.

People with strong attraction or attachment needs can often find a life-style that works for them. The extreme attraction types will fuss over their appearance and hop from affair to affair, while most attachment types will eventually find a partner to fuss over who can make them feel secure. The people who may run into the most trouble seem to be those with very strong needs in both areas, and who therefore want the excitement that comes with newness and instability at the same time they crave the security of an established relationship.

All of us have to balance our cravings for excitement with our needs for security as we make decisions about our relationships. But for people with overwhelming needs in both areas it can be a real problem. We will discuss more extreme situations later, but first let's look at the normal problems of staying in love.

THE DIFFICULTIES OF
STAYING IN LOVE

In our society the classic pattern for a lasting romance to follow is for two people to start off feeling attracted to each other, and then, over time, allowing their attachment to grow. A sense of excitement brings them together, but over a span of years a sense of security and mutual comfort becomes their most important link. Many lasting attachments do occur without strong initial attraction, such as between friends or otherwise well-matched individuals (as in arranged marriages). Other marriages, without much attachment, survive for reasons such as convenience or a commitment to raising children. But these are usually considered compromises or less desirable variations on the classic theme as far as our society is concerned.

Unfortunately, for a variety of reasons many people are having trouble following the classic pattern these days. Those whose attraction or excitement needs are too strong often don't make successful marriages, because they can't give up seeking new attractions. From the opposite extreme, people whose attachment needs are overpowering may demand so much commitment from a partner so quickly that they scare people off, unless they happen to meet a like-minded soul.

Supposing, however, that your attraction and attachment needs are fairly normal, and you meet someone you like, fall in love, maybe live together for a while, and eventually get married. What happens next? This is something you can't find out from going to the movies, because what happens next is not dealt with in films. Nor is it dealt with in popular music or popular novels or very much anyplace else.

The problem is that in becoming attached to one person, and then cementing that attachment (usually by marriage, but it could also be by simply living together), you have traded off some excitement for comfort and stability. In essence you have given up the thrills of the chase for the security of a committed relationship. The problem then becomes: How do you keep it from getting dull?

This is an issue that all but the most security-minded among us, who don't care about excitement, must face. Some very well matched lovers say that their relationship gets more exciting with time. Others may not feel that being married means giving up other lovers. This is one way out of the bind, but a risky one; often either the affairs or the marriage are eventually given up.

Most people, however, are aware that they have made a trade-off. If it feels worth it most of the time, they stick with the marriage; if it doesn't, they don't.

I deliberately said "worth it most of the time" because few married people that I know, no matter how generally satisfied they feel, experience their marriages as "worth it" all the time. Every relationship has its rough spots, and our memories help get us over these. Have a fight with your wife and you may suddenly feel: "Why do I live with this monster?" Then you remember some of the good times you've had together and you know why. The key is being able to recall these nice experiences even when you're very angry. This tempers or balances the anger and kindles the hope that more good feelings lie ahead. It gives you a reason to stick around when your feelings of attachment are being strained to the utmost.

The importance of this is well known to the lovers, friends, and therapists of people who lack this capacity for remembering the good times when things go bad, and instead get totally caught up in their immediate feelings. If they are angry, they feel enraged and cannot summon up any counterbalancing memories. Similarly, if they feel hurt, the person who hurt them is a total, absolute creep and has always been one. If in fact the partner always *does* treat the person badly, then it's useful to realize this. But if not, if instead the past is being falsely and negatively colored by the hurts of the moment, then it's a problem.

It's a problem because people overwhelmed by or prey to their immediate feelings tend to act out on those feelings. Acts of violence to the partner (more common for men), harm to oneself (more common for women), or walking out are what we usually see. One should not be too forgiving if your partner really does treat you badly much of the time. But relationships require the ability to balance the difficulties of the moment with an appreciation of the overall value of the tie. If you can't do this, you can't sustain a relationship, unless you're willing to stick around just because you're too terrified to leave.

Above and beyond sustaining the tie, however, the trick is to have some fun also. This book is not a marriage manual, so we can't go into how to solve the eighty-four million difficulties that couples encounter. But the biological principles we have talked about so far provide an insight into some of the things successful couples intuitively have arrived at for themselves.

The most important is that you need to provide situations where both you and your spouse get your pleasure centers stimulated in and by each other's company. This is what keeps attraction alive, so you don't have to rely solely on your attachment (security) needs to keep things together. You need to be romantic at times. You need to go out to candlelight dinners. You need to take vacations, no matter how brief, from work, the kids, or the house. You need to have fun together. You need to set aside times where you can relax together. They don't have to be once a day, and shouldn't be once an hour; but also they shouldn't be too infrequent.

You also need to do things that break down barriers between you and your partner, because people seem to be wired to find this very heady. Mutually pleasurable sex is great for this. So is a shared hobby, passion, or interest. Especially helpful is anything that conveys your concern and wish to see your partner grow or succeed as a person.

Being able to problem-solve together is also important, unless one of you is always willing to take the rap when things go wrong. Most people aren't, so anger lingers unless each of you can both see and admit when he or she is wrong. Fights get ended the quickest and with the fewest hard feelings this way. In fact, getting over any fight, i.e., really making up, can be a high, because you are suddenly hit with a big change in your level of intimacy. That's why it's important not to do or say anything during a fight that will leave a lingering hurt.

There are many reasons why so many people are getting divorced these days, not all of them bad. But many people getting married are simply unprepared for the experience. They have the idea that the relationship should always be exciting, which it can't be. At the same time, they don't know how to maximize the possible excitement they can have. Most have little idea how to work out difficulties or disagreements. Divorce is easy; so are affairs. Perhaps the really interesting statistic is how many couples manage to make it rather than how many don't.

Professor John Money of Johns Hopkins Medical School is one of the few American medical researchers who has specialized in the biological aspects of what goes on between human partners or couples, a field that he calls sexology. In his most recent book, *Love and Love Sickness,* Professor Money outlines his views on the prerequisites for a "long-lasting reciprocal love match." The first requirement is romantic attraction, which involves finding someone suitable for you to project an idealized image of your own onto. Professor Money calls these projected images "love blots" because the process is similar to what is done with Rorschach inkblots.

The second requirement for a successful relationship is that your love blot not be too far from reality. As Professor Money puts it, there should be "a very close fit between the actuality of each partner and the love-blot image projected on to him or her by the other partner, and this is a two-way fit."

The third requirement for a lasting romance is that if a person's idealized images and expectations change, a complementary or mutual process occur in the spouse, so that the couple grows together rather than apart. Unfortunately, there is yet no foolproof way to make this happen.

I am more optimistic than Professor Money about what couples can do to keep their romances alive. He believes that a couple's passion for each other is typically at its peak for two or three years, and that this may be nature's way of ensuring that a pregnancy occurs. While we can't make people stay in love, I believe there are many things we can do to keep ourselves and the partners who start off loving us from falling out of love. As I have tried to indicate, romantic love involves our basic neuro-chemical systems. Like most other systems, the better you understand them, by and large the better they will work.

Now we have another way to understand why some relationships that lack strong initial attraction can still be successful over time, and others that have this attraction may not be. Strong initial attraction often involves idealizing, which may or may not hold up over time as you get to know your partner better. On the other hand, falling in love slowly, as with someone you work with or have become friends with over time, is less likely to be filled with unpleasant surprises, because you know the person better. Finally, marriage partners picked, either by you or by a matchmaker, because of your common backgrounds, interests, and so forth, may have the greatest likelihood of changing with time in directions well matched to your own development. Thus while strong initial attraction most closely fits our notions of romance, it may also lack some of the long-range advantages found with its less passionate but perhaps equally durable alternatives.

PROBLEMS IN LOVING

A variety of common but highly troubling romantic difficulties make more sense when examined in terms of the attraction and attachment mechanisms we discussed earlier. These include clinging and loneliness, instant bonding, unrequited love, and delusional loving.

Clinging and Loneliness

Some otherwise puzzling emotional patterns could be caused by unusually strong needs to feel attached or to intense fears of separation. These include the tendency to jump hastily into relationships; the compulsive need to socialize; more than usual difficulties with loneliness; and too much dependency or submissiveness in what looks to everyone else like a bad relationship.

According to some recent studies, human infants manifest their attachment to their mothers by the extent of anxiety and unhappiness they experience when they feel abandoned. In adults who find themselves romantically unattached, these same feelings sometimes lead to a tendency to jump into relationships hastily. Such people feel they have to have a partner, and are often not very particular except that the partner not expect them to be independent. I have one patient, a female attorney, who has made three hasty marital choices, and they've all worked out badly. She feels an intense need to have a man support her financially. The only problem is that she also can't stand to be alone so she doesn't take the time really to screen the men she gets involved with. And each of the men she has married has wanted her to support him.

Another pattern, engaged in by people both with and without primary attachments, is to socialize compulsively. This is different from simply being outgoing or gregarious and having a wide circle of friends. Compulsive socialization is a manifestation of a need to feel part of what is going on all the time and has been called "social loneliness." What affected people experience, according to one study, is an uncomfortable feeling of being detached from the community around them, coupled with the sense that whatever they are doing at a given moment is less meaningful or important than what is going on somewhere else. Hence the need to always try to be "where it's happening."

Social loneliness is different from emotional loneliness, according to psychologist Robert S. Weiss. Both social and emotional loneliness are caused by the inability to satisfy our biological needs for attachment. But emotional loneliness involves the need for someone to be intimate with — a spouse, a lover, or at least a very close friend. The symptoms of emotional loneliness are said to include feeling tense, hypervigilant, restless, and chronically anxious as well as loss of appetite and difficulty falling asleep.

Most of us feel lonely on occasion, although we may be embarrassed to admit it for fear of being thought unpopular or unable to make friends. In fact some people are often lonely because of difficulty forming close social relationships. But others, with wide networks of friends, are still excessively troubled by feelings of loneliness whenever they find themselves alone; in such instances, one is tempted to speculate, these people go into a form of "attachment withdrawal" whenever they find themselves alone for more than a few minutes.

Very strong attachment needs can also cause someone to stay in what everyone else would consider to be a terrible relationship. Why some people put up with numerous infidelities, frequent episodes of verbal abuse, lack of emotional or financial support, and even physical mistreatment from their partners has long puzzled many mental health professionals. Such people are often called masochistic, meaning that they in some way have a need to suffer. While this may be true for some, others get stuck in bad relationships because of a fear of being alone. "However bad this relationship is," they reason, "it could be worse if I were on my own." This fear inhibits their assertiveness and creates excessive feelings of dependency. They become unable either to stand up to or to leave the abusive partner.

What creates excessive attachment needs and the crippling emotional patterns that result? Early childhood traumas could play a role. These might include early experiences of abandonment by, separation from, or loss of a key parental figure, which can leave a lifelong emotional scar — and vulnerability — in this area. A lack of parental support for early exploration or independence, combined with the message that one must have someone to cling to in order to survive, could also cause excessive attachment needs. What we see in people with these problems are chronic feelings of incompetence and low self-esteem, and the belief that one cannot make it on one's own.

Up to this point in our discussion of strong attachment needs we have not broken any new ground. Psychologists and psychotherapists have long recognized that some people are troubled by excessive dependency, and many therapeutic hours have been devoted to this problem. Cultural patterns have also been blamed, and a valid argument has been made that females in our culture have traditionally been encouraged to be less independent than males.

But we are in a position to carry this discussion further by looking at the problem from a biological perspective. It is at least hypothetically possible that some people turn out to be more than usually lonely or dependent not primarily because of social or psychological factors but because they were born with unusually strong biological attachment drives. It certainly seems reasonable that people might vary from birth in the strength of this innate mechanism, just as they vary in height or how placid or irritable they tend to be. Many mothers can in fact recall that one of their children tended to cry less when left alone as an infant or was more independent and less clinging as a toddler.

Whatever the original cause of someone's excessive dependency or attachment needs, we know that these feelings operate through neurochemical mechanisms. Whether the actual cause lies in cultural bias, psychological conditioning, biological inheritance, or some combination of the three, for anyone to feel lonely or dependent or fearful requires that certain anxiety circuits go off and pleasure center stimulation decreases when he or she feels alone. While we don't yet have anything like a complete understanding of how these systems work, we do have some evidence that implicates specific areas of the brain as the possible location of the problems and specific chemical systems as potential culprits. What I am suggesting is that whenever we feel lonely or abandoned by or separated from people we are close to, our brains go into a form of attachment withdrawal or craving. Then we need to seek out some kind of social interaction to give us another shot of the chemicals that our brains put out when we are with people with whom we feel comfortable. If this model is right, then it's this chemical outpouring that is responsible for the feelings of comfort we derive from being with friends or loved ones.

The usefulness of this view is that for people with excessive problems in these areas, some form of drug treatment might make sense. I can hear the screams of protest already: "Women have been brainwashed into dependency and you want to give them drugs when what they need is consciousness-raising"; "People have been conditioned or traumatized into feeling helpless; they need to understand and change the way they think about or view themselves through psychotherapy, not to pop some pills."

Two points need to be made. The first is that I am not advocating drug treatment for normal human anxiety or loneliness. As I've tried to indicate, these reactions seem to be part of our evolutionary heritage, and are useful in keeping us involved in established relationships as well as making us seek out new ones. It would not be useful to blunt these with artifical chemicals. What it is useful to blunt is the excessive and overwhelming terror and panic some people suffer when they are alone or on their own. Physical pain provides a useful comparison: to feel none would leave us without a warning system against bodily harm, while intense pain is something we try to alleviate any way we can.

Second, I have nothing against consciousness-raising or psychotherapy, provided that they work. But like any other procedure, they don't always work. There is no reason for antagonism between consciousness-raising or psychotherapy and drugs. The former help with some problems, the latter with others. If you push one cure for every disease, it's not scientific medicine, it's cultism.

If someone has a faulty attachment or separation-anxiety control mechanism, then consciousness-raising or psychotherapy may not be the most efficient way of trying to correct it. It is also possible that some people may be so damaged, whatever the cause, that the panic or depression they feel when confronted with being alone becomes overwhelming. Such people may need pharmacological treatment.

The specific nature of such treatment can be described only tentatively, because no one has systematically tried to give medications to adults with what I am calling attachment problems. However, school-phobic children, the majority of whom really suffer from anxiety over being away from their mothers, do benefit from the same antidepressants that also help adults who suffer panic attacks. It therefore seems reasonable that certain available antidepressant or antipanic drugs might be useful, at least whenever reactions to separation are complicated by depressive or panic reactions. One might also speculate that narcotics or endorphins could be helpful, but it's not clear to me how to overcome the addictive risks that block our ability to study the narcotics, and endorphins are still in very short supply. However, several new narcotic preparations are being touted as nonaddictive. Should this prove to be the case, then such studies could be undertaken.

In the absence of systematic studies we fall back on clinical experience. In the introductory chapter I described an early patient of mine who had suffered repeated depression and appeared submissively stuck in an unpleasant relationship. In her case, treatment with antidepressants not only lifted her depression, but also seemed to allay her fears of being on her won, thereby allowing her to confront her philandering husband more appropriately.

I've recently treated another woman, the wife of a banker, who, despite years of psychotherapy, could not stand up to her domineering husband. Because she reported feelings of depression going back many years, I started her on an antidepressant. Not only did her mood pick up within a couple of weeks, but she also became able to hold her own much better in disagreements with her husband. It seemed that as her depression lifted her feelings of self-confidence and self-worth also picked up.

At this point, we don't have a chemical cure for human loneliness, and I'm not sure we'd want to use it if we did. Loneliness is an incentive to seek out people and form relationships — damping this down chemically would diminish that incentive and give rise to a lot of content but isolated people. Also, most of our chemical cures work on deranged regulatory mechanisms, while loneliness per se appears to be a normal human emotion. But we do have some new treatments that appear promising, at least for attachment or separation difficulties that are associated with panic attacks or depression. Most important, we are only just beginning to scratch the surface in this area.

The Problems of Instant Bonding

This is a problem that has received more scientific study, and the chapter that follows, on hysteroid dysphoria, goes into detail about my research with patients who suffer from it. Briefly, these are people, often but not exclusively female (why this sex difference occurs will be discussed in the next chapter), who get very high and giddy when they begin a romance, become intensely depressed if they feel even slightly rejected, and crave relationships because they usually feel lousy except when romantically involved. On top of all this their choice in partners is usually terrible: the people they tend to fall for were aptly described by one patient as "the flashy, promise-you-anything types." Making things worse, these people are often so demanding of attention and reassurance that even a suitable boyfriend or girlfriend may eventually be driven away.

What seems to be going on with these folks is a combination of unstable pleasure centers coupled with an ability to feel very strongly bonded very quickly, usually to elusive partners. When not involved romantically, or seeking a new relationship, they seem to "idle" at too low a level and feel somewhat lethargic and depressed. When they meet someone they like, however, they get a tremendous rush, so that they become elated and giddy. For reasons that are not terribly clear but which may relate to their intense idealizing of partners, they then proceed to feel very attached, although not in a stable way, since the moment a new partner ceases to be interesting, he or she is quickly dumped. Anyone over the age of fifteen who is ready to marry a person he or she met the day before is probably a hysteroid dysphoric.

The results are predictable, especially since the partners they pick are themselves often highly romantic types who do not get quickly or deeply attached. Given the lopsided relationships that evolve because of the uneven degree of attachment, the hysteroid dysphoric sooner or later begins to feel rejected. Then he or she becomes depressed and often self-destructive.

Why these people pick the type of partners they do is something my colleagues and I have spent a lot of time thinking about. Highly romantic types who will throw themselves into a relationship want someone who will do the same. The kind of partners available to hysteroids are often married or somewhat psychopathic. Furthermore, should they happen to have hit upon a decent partner who is available for a relationship, my hysteroid dysphoric patients often start to get bored or restless. It's as though they can't tolerate someone getting too attached to them, and they often end up dumping this kind of partner. Thus the pattern is one of repeated intense involvements followed by rejection, whereupon they feel devastated.

How do we explain this self-defeating pattern as other than the working of unconscious neurotic mechanisms? Psychological conflict about romantic success may in fact play a part, as may early childhood experiences with emotionally distant mothers and seductive, menacing fathers. But I have also seen patients who started going out with different types of people after successful drug therapy. Several young women used to get involved only with flashy types who would always remain cool or aloof, or with men who already had other serious involvements. This did not change after extensive psychotherapy, but did shortly after beginning a particular type of antidepressant. As a result they are now in fairly stable relationships with warmer and more affectionate men.

Unrequited Love

Even the ancient Greeks knew how to recognize unrequited love. In "The Diagnosis of Love-Sickness: Experimental Physiology Without the Polygraph," Marek-Masel Mesulam and

Jon Perry relate a story about Seleucus, one of Alexander the Great's generals. Some time after the death of his first wife, Seleucus married a woman named Stratonice. Unfortunately, his son Antiochus also fell in love with Stratonice. Recognizing that his feelings could never be returned by his new stepmother, Antiochus resolved to hide his love. He then fell gravely ill.

Seleucus called in one doctor after another, none of whom could discern what was wrong with Antiochus. Finally a celebrated physician named Erasistratos was brought in on the case. Finding no evidence of physical disease, Erasistratos soon began to suspect an emotional problem. By observing his patient day after day, Erasistratos discerned that whenever Stratonice entered the room Antiochus' appearance and behavior would alter markedly. The young man would begin to blush, stammer, sweat, and have palpitations. Equally important, Erasistratos observed that none of this happened with any other visitor, from which he concluded that Antiochus must have fallen in love with Stratonice but wished to conceal this from the world. The story ends happily: Seleucus decided to give up his new bride to save his son's life.

According to psychologist Dorothy Tennov, author of *Love and Limerance*, a surprising number of romantic passions don't end so happily. Impressed with the frequency with which her students were having their hearts broken, she began seriously to investigate what was making these otherwise rational people fall hopelessly in love. The conclusion she came to is that many, although not all, human beings are vulnerable to involuntary states of romantic passion which Dr. Tennov calls limerance. Limerance, or "being in love," is a far stronger emotion than simply loving someone. What makes it a problem, according to Dr. Tennov, is that if we are one of those who is wired for limerance, we may have no control over when or with whom we become limerant.

The problem is that limerance can be willing and wanted or unwilling and unwanted. If you are a limerant type and you meet someone you find attractive, some physical process may get kicked off that is soon beyond your control. You begin to think obsessively about the new person and scan every aspect of your interaction for some sign of reciprocated feeling. At its peak you are a basket case, at least as far as doing or thinking about anything beside being with your LO (limerant object, the person you are in love with) is concerned. With a bad case of limerance you respond like Antiochus whenever your LO comes into view: your heart flip-flops all over the place, you tremble, flush, look pale, and lose all the poise you have worked to acquire.

While in some cases limerance ends in a few weeks, in others it can last a lifetime.

Dr. Tennov believes that either you are wired for limerance or you aren't, which I think is a bit rigid since we see people showing all degrees of attraction and attachment. If we consider limerance to be a form of strong romantic attraction followed by the development of a powerful attachment, then perhaps we have some leads on the actual mechanisms that mediate it. The ideal mix of hope and uncertainty that Dr. Tennov believes is required for limerance to peak makes sense biologically as a way to maximize the emotional ramps we experience (the "changes" we go through) as we oscillate between ecstasy and despair.

The crucial distinction, I believe, is between successful and unsuccessful limerance. Successful limerance, by my definition, is a relationship in which there is a high degree of mutual involvement and a resolution, be it termination or evolution into affection, that is more or less mutually agreed on. Unsuccessful limerance, on the other hand, occurs when love is unreciprocated or only partially returned from the beginning, or where one partner loses interest and the other doesn't. If you have experienced unsuccessful limerance, as almost everyone has, another important distinction is whether you have a single or occasional episode, which may be par for the course, or whether it keeps happening over and over again, which to me indicates that something is seriously amiss and that help is required. While the traditional remedy tends to be psychotherapy, for some of these problems medication can be extremely helpful.

Delusional Loving: A Special Form of Self-Stimulation

I have seen or heard about several patients who said they were in love with their former therapists. Moreover, they were convinced that the therapists, who I knew to be happily married, were secretly in love with them. These patients tended to have other more far-fetched beliefs as well, all of which (including the romantic ones) went away after several weeks on medication.

The psychiatric literature contains several dozen examples of a very extreme form of romantic self-deception which has been termed delusional loving. What this involves is a person, usually a woman, developing an unfounded but also unshakable conviction that some powerful or attractive man is in love with her. Men, traditionally more into power or glory, have been more likely to think they are Napoleon or that they've found a cure for cancer when they become delusional. However, this may change as sexual stereotyping breaks down, John Hinckley's imagined relationship with actress Jodie Foster being a case in point.

There is actually a spectrum of romantic delusions that seem to occur, running the gamut from the far-out to the mildly unrealistic. The most far-fetched is the phantom lover syndrome in which the purported lover is someone who never actually existed. A variation is to feel that one is loved by someone who died or moved away but whose death or departure is denied. A third possibility is to believe (not dream) that someone you have never met, such as a famous television star, is madly in love with you, and is communicating his affection in ways only you can recognize. Moving slightly closer to reality is the belief that some person with whom one has a passing acquaintance, such as a teacher or boss, is secretly enamored. Finally there occurs the more frequent and understandable distortions of sincere but basically nonromantic interest such as might occur in a doctor-patient relationship (which unfortunately on occasions do become romantic).

The emotional dynamics of these make-believe affairs are easily understood. To compensate for loneliness, low self-esteem, or real inhibitions about sex or intimacy, someone begins to imagine that a highly attractive figure, either made up or real, is harboring a great passion. What the deluded lover is doing, in essence, is creating a myth that is highly stimulating to his or her brain pleasure centers. Where the delusional lover differs from the rest of us who use our fantasies in a similar way is by believing that his or her fantasies are real. This probably makes the fantasies much more effective chemically. Since delusional lovers already think they have one, they don't feel the need to actually go out and find a real relationship, which even the most die-hard fantasizers do feel from time to time. Where things can get a little messy, however, is that people nurturing such delusional romantic beliefs sometimes decide to act on them by contacting their supposed lovers. When repeatedly spurned, things may take a nasty turn, including attacks on the unresponsive person, that person's family, or, as in the most notorious recent case, a public figure.

But lots of people feel lonely, unloved, or even unlovable and yet don't totally fabricate a romance, so something else must be involved. The crucial ingredient in delusional loving is the ability to distort reality in the service of emotional needs. Do this to a moderate degree and you get called neurotic or hysterical. But people who totally fabricate romances, especially if their beliefs persist for years on end, have really lost touch with reality. Such people are said to have developed a psychosis, which can vary in severity from an otherwise nonpsychotic individual with an isolated romantic delusion to someone who for years has heard nonexistent voices or had other false beliefs that warrant a diagnosis of schizophrenia.

Psychotic vulnerabilities are based on faulty neurochemistry. Current thinking about schizophrenia, for example, is that the illness involves excessive activity of a particular neurotransmitter system, which fits with the observation that drugs that block this neurotransmitter (dopamine) are effective for treating psychoses. But no one has ever systematically tried to treat delusional romantic beliefs with antipsychotic medication, so it is not possible to say how helpful they would be for this condition.

BIOLOGICAL DISSECTION OF ROMANCE

The idea of using a drug as a dissecting tool may seem a bit odd, unless you try to think of it as a sort of biochemical scalpel or probe. A good example occurred in a large drug study of the early 1960s which compared an antidepressant, an antipsychotic, and a placebo in hospitalized psychiatric patients who were not responding to psychotherapy. Not surprisingly, people suffering psychoses did better on the antipsychotic than on the antidepressant, while the reverse was true for patients who were depressed. But a third group, consisting of people who suffered repeated episodes of panic as well as more general nervousness, was unexpectedly also helped by the antidepressant, even though these patients were not particularly depressed. On this drug they stopped experiencing the anxiety attacks even though they still felt nervous, while the antipsychotic made them worse. Another group of patients with personality disorders did not benefit from any of the medications.

Studying these patterns, the researchers were able to dissect out a new diagnostic group, the panic patients, whose anxiety attacks had been stopped by the antidepressant even though they did not show symptoms of depression. Carrying the analysis one step farther, they also concluded that panic was biochemically different from plain nervousness, since the panic symptoms responded to the antidepressant while the accompanying nervousness did not. If panic was only a more intense form of nervousness or anxiety, as was thought to be the case at that time, then the antidepressant, which helped the severe panic symptoms, should even more effectively relieve milder nervousness, which was not the case.

These principles of biological dissection should be kept in mind as we consider drug effects on romantic love. Lacking systematic studies, we can only reach tentative conclusions and then suggest the additional research that could help us to better understand the chemistry of love.

As I've said before, drugs (or any process) that lower our pleasure thresholds can temporarily make anything we encounter seem more attractive. The problem is that these effects wear off, so someone you've met when your pleasure threshold is low may not seem so appealing on the second date, unless you're at that same level. On the other hand, things that relieve your anxiety, such as a mild tranquillizer or a drink, may make it possible for your pleasure centers to react more freely, which is why liquor and marijuana are such popular social lubricants.

One would not expect the usual antidepressants (such as Elavil or Tofranil) to have any effect on normal romance since they don't appear to affect normal mood or behavior. Where they should have effects are in alleviating people who have become socially or emotionally withdrawn because they can't experience any type of pleasure, in blocking post-romance depressions in certain types of vulnerable people, and in reducing the anxiety over separation in the panic prone.

For individuals with certain other symptom or behavior patterns, a different type of antidepressant (the MAO inhibitors) are dynamite. While these drugs don't have major effects on normal mood or energy patterns, they can, in certain types of people, reduce cravings for romance, allow people to feel good about themselves even when unattached, help overcome patterns of romantic avoidance that often occur in very sensitive types who have their hearts broken, and even foster more sensible and potentially successful choices about with whom one gets romantically involved.

Several available antidepressants are very effective at blocking panic attacks. If panic anxiety is related chemically to excessive attachments, loneliness, or reactions to separation, perhaps the drugs can help in those conditions as well. This is something that needs a lot more study.

Because we think that endorphin systems play a role in this area, it would be very exciting to examine morphine's effects on our feelings of attachment or reactions to separation. However, morphine and related narcotics are highly addictive, so it's a problem to study them. What we can do, however, is look at the effects of nonaddictive narcotic blockers on romantic attraction and attachment. Any changes we see with such agents can be ascribed to their ability to counteract the effects of naturally circulating endorphins.

My prediction would be that a narcotics blocker would not affect our feelings of excitement on meeting someone new, because our feelings of romantic excitement do not involve our internal narcotics (endorphin) system. What a narcotics blocker should blunt, however, are the feelings of satisfaction we experience with familiar loved ones or even close friends, since I believe the warm feelings we get in these situations are caused by stimulation of our internal brain narcotic system.

Lithium, on the other hand, might very well affect our feelings or romantic excitement.

We know that lithium, which chemically is very similar to table salt, blocks mania as well as amphetamine- and cocaine-induced euphoria. All of these are thought to involve shifts in neurotransmitter activity. If a similar process is involved in romantic highs, then lithium might have an effect here as well. Many patients on lithium still maintain satisfying intimate relationships, so clearly the drug does not totally block romantic relatedness. But if massive brain stimulant shifts are in fact part of what goes on when people feel swept off their feet, one might expect lithium to take the edge off this, as it did with one patient I described earlier. At the same time, lithium should leave our attachment feelings, which don't involve the same brain stimulant changes, and which play a bigger role in established relationships, undisturbed.

In one study, which unfortunately still must be considered inconclusive, the reactions of normal men to two weeks of lithium and two weeks of placebo were compared. The drugs were administered "double-blind," which means that neither the researchers nor the subjects were supposed to know who was getting what drug at any particular time. (I say "supposed to" because lithium produces side effects, so it's possible that both experimenters and subjects could make some educated guesses.) When on the lithium, the research subjects felt less inclined to deal with other people or respond to new stimuli. Also, while on the lithium they rated their desire for excitement, attention, physical contact, and just being with people as less than when on placebo.

The changes induced by lithium were not very strong, and may have become even less with time had the research subjects had more of a chance to become used to it. Nevertheless, the results do suggest that lithium can exert some influence on the amount of excitement normal people derive from social encounters or novel stimuli, both of which are aspects of romantic excitement.

Drug treatment can also make the emotional dynamics of a relationship much clearer. When I first began my psychiatric practice, I naively assumed that improvement in the way people felt or functioned would automatically benefit their close relationships. Surprisingly, this is not always the case.

What opened my eyes were several instances in which patients became dramatically better with treatment, only to have their marriages come apart at the seams. One patient was a thirty-year-old man with ten years of depression whose wife appeared desperate to see him well. After six weeks on the right drug he was a lot better. And what did she do but take off. Either this woman unconsciously preferred her husband when sick rather than well, or she was fed up but too guilty to leave until he was well enough to fend for himself.

A similar situation involved a woman with recurrent depressive episodes whose marriage is very close to the rocks if not on it since I helped stabilize her with medication. Turns out that she and her husband get along fine when she is sick and he can nurse her, which is the way they originally came together. When she is well, these roles are no longer available, and they can't get along at all.

The bottom line is that when you treat someone with psychoactive drugs, you may do more than alter his or her internal chemical balance. What may also be altered, in ways that are sometimes surprising or unpredictable, is the power balance in the intimate social network. People get used to an individual's being and responding in certain predictable ways. For people with chronic anxiety or depressive conditions, drug treatments can sometimes change this dramatically, usually in the direction of reducing dependence and increasing assertiveness. This in turn demands a big adjustment from the intimate partners, family, and friends, who usually welcome the changes. But for relationships that depend on one or both partners being in some way excessively vulnerable or unstable, sudden changes toward health in one of the participants can upset things. This can leave the other wishing these changes had never occurred, or even actively working to make them disappear.

Understanding the chemistry of romance and its problems should make things clearer for many people whose love lives are "normal" as well as for those who are having trouble. For some of us, however, understanding will not be enough. The biggest risk we all face when we get involved romantically is of getting our hearts broken. While there is neither a preventive nor a perfect cure for this, a lot more is known than most people realize. This is what the next chapter is all about.

CHAPTER 6

Broken Attachments and Broken Hearts

Everyone is saddened and hurt to some degree by the breakup or end of a romance, involving as it does the loss of intimacy, a need to start over, and often some sense of failure, rejection, or guilt. Given such an experience, it is natural to go through a period where one feels a sense of loss, grief, sadness, and mourning. One may also feel hurt, bitter, and distrustful, as well as fearful of ever getting so involved with anyone again.

Fortunately, for most people these feelings pass with time. Exactly how much time seems to depend on a number of factors. These include the length and intensity of the broken relationship; our prior experience with loss and separation as well as future prospects for romance; the particulars of the broken tie, especially the way it ended; certain biological and psychological factors that together determine our natural resiliency and ability to bounce back after disappointments; and the degree to which our need for social support and temporary props to self-esteem can be met by family, friends, job, and other elements of our lives.

However, many people seem to have an especially hard time of it after a romance ends. This usually takes one of several forms. Some become wildly promiscuous, dating and having sex willy-nilly in a misguided attempt to repair their self-esteem. They hit the bars and discos with a vengeance, picking up, sleeping with, and then discarding a different man or woman each night. Not at all ready for another relationship, they apparently have special motivations: a search for both sexual pleasure and closeness through sex; a need to prove that they are still attractive to others; and a wish to control and discard others as they feel they had been controlled or discarded. With time and a chance to reflect, this phase usually passes.

A second unfortunate outcome at the termination of a relationship is that someone may become severely depressed. This can be distinguished from the normal sadness we feel if someone has left or grief if they have died. Normal loss or grief involves sadness, crying, anger, preoccupation with the lost partner, a temporarily reduced enthusiasm for other aspects of one's life, and often some initial disturbance in appetite or sleep. Such feelings usually subside by the end of three to six months. Depression, on the other hand, is characterized by a more intense or protracted state that can also include profound hopelessness about the future, suicidal wishes or acts, and more severe disruption of normal social life, work routine, or activities such as eating and sleeping.

Why some people become severely depressed at such times is not clear. At least in some cases more general vulnerabilities to depression are probably involved, but where this comes from is uncertain. Some vulnerabilities appear inherited, since the children of depressed people also show a high rate of depression. Countering those who view this as solely a matter of environment (if you're brought up by a depressed person you will "learn" depression), it has been shown to be true even for those people who were given up for an adoption at an early age and raised by nondepressive adoptive parents.

Certain early life experiences may create a later tendency

toward depression. Losing a mother at any age before fifteen, or a father between ten and fifteen (especially for girls), seems related to later depressions. Whether early losses actually create this vulnerability or only worsen a preexisting inherited problem is unclear. Perhaps later events, such as a romantic breakup, have a greater impact on people who have also lost a parent early in life, because of the meaning those people place on it, i.e., the loss of a lover reawakens the thoughts, feelings, and perhaps the neurochemical reactions, of the earlier, even more painful life event. But many adults who become depressed after broken romances do not have histories of childhood separation or loss, so other factors must also be involved.

The treatment for someone who has become seriously depressed following a romantic loss depends on the specific pattern of symptoms. The first thing we do at the New York State Psychiatric Institute's Depression Evaluation Service is to take a detailed clinical history. For the treatment to be helpful I need to learn how depressed the person is, if there is a risk of suicide or other self-destructive behavior, and if the person is suffering from any of the biological changes that often accompany depression. I also ask if there is some pattern of previous episodes, and to what degree the person contributed to his difficulties, as a way of determining what, if anything, he needs to do differently in the future to avoid similar occurrences.

In terms of immediately helping someone, I will first try to establish a sympathetic and supportive doctor-patient bond, emphasizing that virtually all people suffering depression can be helped to feel better (and that many get better simply as a result of time). I will also look for signs of altered brain chemistry to see if one or another type of antidepressant medication should be part of the treatment. For people whose pleasure centers have completely shut down, so that they are clearly melancholic (where they don't desire food, have lost interest in sex, can't stay asleep, and enjoy nothing), standard antidepressants are very helpful. For depressions that show the reverse pattern — with oversleeping, overeating, extreme fatigue and the ability still to be cheered up — our research strongly suggests that another, little-used type of medication is more helpful. Just as for depressions that are not related to romance, antidepressant benefits usually take two to four weeks to appear, at which time mood, energy, appetite, and sleep begin to normalize. There may still be sadness over the specific loss, but it no longer incapacitates the person or casts such a pall over his or her view of what lies ahead.

Many people who seek help after a broken relationship report feeling very depressed but do not have prominent disturbances of sleep, appetite, or energy. Such people usually complain of intense sadness, frequent crying, inability to concentrate, loss of interest in work or daily tasks, and preoccupation with their lost love, coupled with despair over ever finding a replacement. Some also seek solace in drugs or alcohol; others have seriously contemplated suicide or have even tried to kill themselves. Such people, whose symptoms date from the time of their disappointment (which is usually at least several months before I see them), but were well before that, do not seem to require specific antidepressant drug therapy to get better. What we find is that they improve whether a drug or a placebo is prescribed. What may therefore be most helpful is a warm and supportive figure who understands what the person is going through and can offer reassurance that with time the terrible feelings will pass. Occasionally, however, people get stuck in one of these "disappointment reactions"; for those whose symptoms persist for a year or more, who seem to be overwhelmingly affected by their depression, antidepressant drug therapy appears helpful.

Some people seem to get very stuck emotionally after a relationship ends. In these cases one of three things may be involved.

One is that a person may have slipped into a chronic depression that involves some long-lasting change in brain chemistry. For reasons that are unclear, time does not lead to self-correction, and the process continues for years unless effective drug treatment is initiated.

A second possibility is that the person has become demoralized. Take the example of a good swimmer who gets caught in a fast current, almost drowns, and then becomes fearful of the water. After a broken romance everyone's confidence will be shaken at least a little. But some people, especially those who also suffer depressions, lose their confidence completely, and so they become convinced that to try again would result only in another failure. They begin to think that there must be something wrong with them, and that no one would find them lovable. The treatment is the same as for our hypothetical swimmer. You have to get them back in the "swim of things," so that they can find out that they can still swim (or love and be loved) without drowning.

People who continue to "pine away" for years on end for a lost love may be doing so because they are blocked from getting involved with someone new. In general I don't believe that there is only one person out there who is just right for us, so that if we lose that opportunity our possibilities for love are over. To feel this is to be closed off to the excitement of new people and new romance. This can come about out of depression or demoralization; it can also come about when people become fearful of romance.

A surprising number of people give up on romance after a few — sometimes even a single — bad experiences. Often such people are highly sensitive to rejection, and get even more than usually hurt when a relationship falls apart. After several disappointments, they begin to avoid anything that hints of romance. They don't go to parties. They won't allow their friends to fix them up, and they push away anyone who begins to express an interest in them. It's as though they've decided "no more romance — it's too dangerous for me." What is turning out to be so interesting is that this excessive sensitivity to rejection seems to have a biological basis, and is part of a larger emotional pattern we call hysteroid dysphoria.

HYSTEROID DYSPHORIA: LIFE ON AN EMOTIONAL ROLLER COASTER

I first heard the term *hysteroid dysphoria* a number of years ago at an international gathering of mental health professionals who had come together to discuss treating patients whose severe emotional difficulties didn't seem to fit traditional diagnostic categories. The "stars" of the meeting were such psychoanalytic luminaries as Otto Kernberg, Margaret Mahler, and Wildred Bion, and their papers were well received by the audience, which numbered over a thousand. However, there was one other paper that generated a certain stir at that meeting, although it seemed that most of the audience didn't really know what to make of it at the time. It was given by an expert in drug therapy named Donald F. Klein, who had been invited to speak on the use of various psychiatric medications for these "emotional" disorders.

Klein used the term *hysteroid dysphoria* to describe one such pattern of difficulty. People with hysteroid dysphoria, he suggested, were individuals whose fundamental problem was their tremendous sensitivity to rejection. Along with this went the need for a great deal of attention, praise, and admiration to maintain their mood, energy, and self-esteem. Major rejections were devastating, and even minor ones seemed to have a profound effect, plunging these people into depression from which they would emerge only when something happened to cheer them up.

But they sure could be cheered up. What it would usually take was attention from the right person. If the boyfriend or girlfriend who had been cool or rejecting the night before would call and sound interested, the terrible despair would suddenly lift, and everything would once again feel fine. If a boss who was aloof one day was warmer the next, the same thing would happen. These effects could be quite dramatic at times — both the sudden crashes into depression as well as the quick lifts back to well-being — and confused everyone: patients, families, friends, and mental health professionals alike.

At this point many of you must be asking yourself: "So someone gets depressed after a romantic breakup, or feels good if the lover with whom they've had a fight calls the next day. What makes this a psychiatric disorder? If it is, aren't we all hysteroid dysphorics?"

What makes the people we call hysteroid dysphorics different from the rest of us is that while love, romance, and attention from others are things that enrich our lives, we can live, function, and at times enjoy ourselves without them. Not so for people with hysteroid dysphoria. For them there is no stable middle ground. They are either in love or the center of some kind of positive attention and feeling up, or they are not getting that kind of attention and are feeling rejected or miserable. Thus the need for attention and sensitivity to rejection become the dominant themes in their lives. The most common pattern that results is one of repeated intense romantic involvements that ultimately, and predictably, don't work out.

We've been running our special treatment programs for several years so I have dozens of case histories in my office files. Steve and Jan are two that come to mind. Steve was in his late thirties when he came for treatment. An energetic and athletic man (when he was feeling okay), he'd gone through a half-dozen serious romances (including one marriage) by the time I saw him. I say "gone through" because the pattern was the same in each one.

When Steve would first meet someone, he'd get very excited. This was probably contagious, because he says his relationships always start off with a bang. For a few months he'd be on top of the world. Then always something would start to go wrong.

What would go wrong was that Steve would start to get very possessive and jealous of anything his new partner did without him. This might be flattering at first, but hard to take after a while. His new girlfriend (in one case wife) would then start to complain. Fights would follow, and Steve would end up severely depressed. (He would stop exercising, stop seeing friends, and find it a struggle to get to work at such times.) He might then control his possessiveness and jealousy for a while, in which case things would pick up. But it would never last.

Jan's story was much the same. Jan was a bright, vivacious woman who had no trouble getting men interested in her. For reasons she could not identify, however, they didn't stay interested. When we went over each relationship in detail, Jan's pattern was much the same as Steve's. As were those of most of the hysteroid dysphoric patients I've seen.

When they first meet someone they are attracted to, hysteroid dysphorics become elated — their energy goes up, they need less sleep, their appetite is diminished — in short they feel "high." Moreover, they are convinced they've met the love of their lives — they call all their friends to tell them, and they are ready to move in, relocate, or get married — all in the first week. The only problem with all this is that their choice of partners is usually terrible. The Mr. or Ms. Right they've just fallen for may be married, out for just a good time, a habitual single, or otherwise unavailable. In short, the partner often has a very different agenda than our subject.

You can all see what's coming — well, all of you except someone with hysteroid dysphoria. What's coming is a big disappointment, or as Jackie Gleason used to say, "Pow, right in the kisser." It may take a week, it may take a month, or it may take a year, but eventually it comes — the Big Rejection. It may be about time together, or living together, or about getting married, or about monogamy, but one way or another, our subject begins to feel rejected, as the partner begins to indicate that he or she is not ready to leave a present marriage, or get into another marriage, or not date anyone else, or whatever.

What happens then is simple — the hysteroid dysphoric is devastated. Not just sad or blue or disappointed, but devastated. Life loses its meaning; work becomes a drag; why see one's friends? The only solaces are bed and chocolate — and long days are spent overdoing both. The highly energetic, enthusiastic individual of last month can barely get from here to there — it's literally as if someone had pulled the plug.

Some of my female hysteroid patients in the corporate world seem to have a particularly hard time of it, because married male executives are always making passes at them. This used to be a real problem for Mollie L., who has been in treatment with me for several years. While most of the women who worked with her understood the rules of the game — that nothing serious will come of these affairs — Mollie didn't seem to learn and kept falling in love. While the men she fell in love with seemed to have cared for her as well, none was really interested in marriage, which is what Mollie had in mind. Thus in the end she'd always be hurt and then depressed.

How long she or other dysphoric patients stay depressed can vary. It can be a day, a week, a month, or six months. Some people manage to drag themselves along despite feeling miserable, others remain attached to their beds, and yet others frantically try to repair the damage by finding someone else.

These latter types hit the bars and discos with a vengeance, and a different partner every night for weeks is not uncommon. All may numb their sorrow with alcohol or "downs," and some have discovered that amphetamines or cocaine give them a temporary lift.

"Well," some of you may still be saying, "that's happened to me once or twice — does that make me a hysteroid dysphoric?" The answer is no! Anything can happen to someone once or twice. But the typical hysteroid dysphoric is a junkie — an "attention junkie." And like other junkies, there is no learning from experience. Thus it doesn't happen just once or twice. *It happens over and over and over again.* Great euphoria in the beginning, devastating depression when it doesn't work out, and little or no ability to understand, let alone change the pattern.

It is this pattern that first tips us off to the problem — after all, many of these people are bright, attractive, and talented. Moreover they are very engaging; I often find myself wishing I'd met this person socially instead of professionally. Yet they are continually getting involved with wrong types — and continually getting hurt or dumped.

There seem to be a couple of variations on this theme. Occasionally, a hysteroid dysphoric will meet someone who isn't married, aloof, or committed to uncommittedness — in short, someone available for a relationship. Well, what happens then? Sometimes things may work out — we don't hear from or see these people because they're happy. But many times things don't work out, and again, the pattern is pretty constant.

People with this kind of problem have trouble feeling satisfied simply by being in a relationship. Rather, they seem to need the kind of attention one usually gets from a partner on a honeymoon — where two people are together all the time and involved in nothing but each other. This is fine on a honeymoon, but more than a little hard to manage after that. In the normal routine of life partners sometimes stay late at work, see other friends by themselves, or happen to roll over and go to sleep right after making love. The hysteroid dysphoric's response to this is to seek reassurance, and "Do you still love me?" becomes an everyday question, usually followed by the accusation "You don't love me anymore, or you wouldn't do all these things to hurt me." What's happening is that when the literal or figurative honeymoon ends, our subject starts to feel rejected. "Why does he [or she] want to play tennis or go bowling with this friend, instead of being with me?" "Why don't we talk after making love anymore?" The "hurts" accumulate, the feeling of rejection deepens, the mood and energy sink, and jealousy, anger, depression, or despair sets in. These are accompanied by tears, rages, threats to leave, threats of suicide — all really demands or pleas for the partner to change. Having picked an attractive but often emotionally distant partner to begin with and/or lacking the patience, skills, and independence to work for emotional involvement, the instigator finds that these maneuvers don't help. Or if they help, they don't help for long, and either the partner is finally driven away or our subject takes another lover to recapture some of the good feelings. Sometimes the primary relationship continues, sometimes it stops and then starts again (and again and again), and sometimes it just stops. Whatever the particular scenario, it's usually a never-ending roller coaster of highs and lows.

Another scenario is a variation on the theme "Once burnt, twice shy." After one or two or half a dozen painful experiences, a person with hysteroid dysphoria may give up — I call it "putting themselves on the shelf." They simply take themselves out of circulation, forswear romance, and begin to avoid romantic intimacy like the plague. Though they still hunger for it, and never feel as good as when they used to be in love, they have decided that it's all simply too painful, that they'll just get hurt in the end anyway, that available partners of the opposite (or sometimes same) sex are all selfish and exploitative anyhow. So why even try?

A number of my patients show this third pattern, and it's tough to change. Faye S. was a former actress in her thirties when she came to see me. She'd had a passionate affair with an older man some years prior which had ended when Faye's lover went back to his wife. Faye had been chronically down in the dumps since, and, you might say, ardently avoided men. Putting her on a particular antidepressant quickly restored her joie de vivre in other ways, but it took a couple of years for her even to think about dating again.

I might add here that these patterns of romantic involvement also get played out by people with hysteroid dysphoria in other areas of their lives, particularly at work. Thus we see bright, talented people who enter jobs enthusiastically, only to "run out of steam" after six months when the attention that new people get on a job begins to lessen. We also see people who roller coaster through work depending on how they're being treated. And we see people who have given up, such as the actor or actress who will no longer audition for fear of rejection.

While descriptions and hypotheses about hysteroid dysphoria are all quite new, the first reference that I have found about people with this type of problem actually occurs in a book called *Mind and Matter*, written by a psychiatrist named J. A. Milligan in 1847. To quote Milligan:

> The nervous temperament as I have already observed, may be considered of a complex character. In this constitution the sentient (feeling) system predominates, and one might say that the frame is all sensation; a vivid susceptibility to all external impressions prevails . . . they are constantly seeking sensual enjoyment and novel excitement. *Love, or what they fancy to be love, is to such*

individuals a necessary pabulum. [emphasis added] For a while the attachment is ardent and enthusiastic; but as selfishness and fickleness are the attributes of this temperament, their affections are changeable, and rarely of long duration; and their vanity once offended, they can hate as fiercely as they adored the former idol of their worship. Their great irritability, both in their moral and physical faculties, will sometimes render such persons miserable; for they are jealous, suspicious, and impatient, and never seeking to ameliorate their condition, they must be subject to frequent disheartening disappointments. Thus, miserable themselves, while in the vain pursuit of an imaginary happiness, they involve in their sufferings those who have had the weakness to rely on their professions, or attach themselves to their checquered destinies, unless they are of a similar temperament, and seek fresh emotions to replace past enjoyments and revive faded pleasures. Females of this constitution are subject to constant hysterical and convulsive affections, that often render them a plague to others and a nuisance to themselves. Their ideas are as romantic as their partialities are whimsical and unaccountable; vivid emotions constitute their life. They must breathe an atmosphere of excitement, or linger and pine away in self-inflicted consumption, or what they fancy a broken heart.

Thus more than a hundred and thirty years ago people with hysteroid dysphoria were recognized, with their addiction to falling in love, their sensitivity to and overreaction to rejection, their jealousy and possessiveness, and their need for constant romantic involvement.

But what have modern psychiatry and psychology had to say about such individuals, or to offer them in the way of help?

Here we can observe a curious thing. In the first half of the twentieth century American psychiatry came to be dominated by the psychoanalytic school of thinking, which tended to view patients with what we call hysteroid dysphoria as severe hysterics. According to Freudian principles, "hysterics" seemed to have classic "oedipal problems" (a colleague's patient used to call herself an "Oedipus Wreck"). What this means is that a person has not really given up on the idea of having the parent of the opposite sex for a partner, as all of us are supposed to do by the time we reach adulthood. Still trying to have Daddy or Mommy for oneself makes us too guilt ridden or anxious to be successful with anyone else. Any other partner is too much of "stand-in." Hence a vivious cycle — we seek partners, turn them on and are turned on to them, and then flee or drive them away when things get too hot or close.

Sounds simple. Well — in psychoanalytic terms it seemed clear, and "hysterics" were considered good candidates for the couch. Often bright and verbal, they could talk about feelings, remember their dreams, have lots of colorful fantasies — certainly they weren't boring to work with.

Unfortunately things don't always work out as planned. Some "hysterics" did better, perhaps by learning how their unconscious wishes and conflicts were messing up their lives. But others got worse, and for a while no one could figure out why.

Let me digress here for a minute and ask you to think about what might happen if you take a person with what we call hysteroid dysphoria and put him or her in psychotherapy for an hour several times a week. What happens is that the patient-therapist relationship becomes a very important *personal* relationship for the patient. And we already know what happens to people with hysteroid dysphoria in their important relationships. They become very romantically involved, very sensitive to criticism, feel devastated by a lack of attention, and get very possessive, jealous, and demanding. And that is exactly what started happening to these people in psychoanalysis and traditional psychotherapy.

The analyst would try to point out such-and-such a pattern — and the patient would feel "He's attacking me"; "He thinks

174

it's all my fault and I'm to blame and I'm no good." The analyst would go away for a vacation and the patient would feel, "If he really cared about me, he wouldn't do that," or "He would take me with him," or whatever. The analyst would say at the end of an hour, "We'll talk about that at our next session" — and the patient would feel: "How cold and cruel and unfeeling to cut me off now." The analyst would say, "We can only have a professional relationship"; and the patient would feel: "If you you really want to help me, you would see how much I need you to love me."

To some degree many patients in psychoanalysis or intensive psychotherapy have these kinds of feelings. But the difference is one of degree — how strongly they feel it and what they do about it. People with hysteroid dysphoria tend to feel it very, very strongly and, more importantly, to do a lot about it. And what they do about it often makes a shambles out of the psychotherapy.

Many disrupted therapies, hospitalizations, overdoses, and theories later, these patients continue to vex the psychiatric profession. This is what led to a discussion of such patients at the international conference I was attending, and to my first encounter with the concept of hysteroid dysphoria.

As I mentioned earlier, hysteroid dysphorics appear to have an underlying biological vulnerability, a biochemical instability that makes them need attention and praise to feel okay, and one that causes rejection to affect them more strongly than it does most people. They also appear to get attached to people very quickly and to protest very loudly if someone they've just started seeing tries to break off with them. This has made me wonder if hysteroid dysphorics don't also have some defect in their attachment mechanisms, creating a tendency toward "instant bonding."

Whether they really do get attached (as opposed to attracted) to their partners is unclear, however, because hysteroid dysphorics often lose interest once their partners appear committed.

Earlier we discussed the biological mechanisms that exist in all human beings and that cause us to feel a mood lift when we are being praised, admired, or applauded, and to feel sad or despondent or down when we are disappointed, criticized, or rejected. How this mechanism works is not totally clear, but it seems to have a chemical basis. That is, some kind of stimulating substance is secreted in greater amounts during socially rewarding encounters, giving us an internal "high." The elation that we feel when we fall in love is a good example. Conversely, the level of this chemical would fall during socially unrewarding encounters, causing us to feel temporarily sad or despondent.

People with hysteroid dysphoria are much less biologically stable in this regard. In particular, they have trouble just feeling okay. If their brain chemistry is not constantly stoked by attention, levels of something seem to fall off very rapidly, leaving the person feeling depressed, lethargic, or worthless. It's as though the system is poorly modulated, easily swinging from one extreme to the other but not very stable in the middle range.

The next logical question is: What is the chemical? We get a clue if we look at the symptoms that accompany mood swings. When depressed, hysteroid dysphorics overeat, oversleep, and feel extremely lethargic. Do you know what happens to someone who takes stimulant drugs such as amphetamines or other diet pills for a while, and then suddenly stops? They crash — they feel depressed, they crave food, they sleep a lot, and they have no energy. The same as someone with hysteroid dysphoria when he or she feels rejected and gets depressed.

Similarly, what is someone like who is "on" amphetamines? He or she is high — talkative, euphoric, full of energy and ideas, often very social, with little desire to eat or need for sleep.

Again, much like a person with hysteroid dysphoria in the initial throes of a romance or when the center of attention at a party.

The parallels are clear, and they suggest some kind of an internal amphetaminelike substance that is stimulated or secreted when people with hysteroid dysphoria have positive interactions, and switched off or used up when they have negative encounters. This would account for the great fluctuations in mood, energy, need for sleep, and appetite, as well as the dramatic shifts in their general feelings of well-being and self-worth that we see in these individuals. It would also account for their "craving" for attention and romantic involvement.

But where would this "brain amphetamine" come from? Well, this is a hard question to answer, as we've only just begun to study the problem. We do have a "suspect" however — the substance called phenylethylamine (PEA) that is found in the human brain in small amounts and closely resembles amphetamine in its chemical structure. The amounts of PEA found in the urine of some depressed patients have been low, but these assays may not have been accurate. PEA has also been shown to have amphetamine effects when administered to monkeys and to mice, suggesting that it could also have amphetaminelike effects on humans. We are presently studying levels of PEA in people with hysteroid dysphoria and others during different mood states. In addition, chocolate is loaded with PEA, and we have observed that many of our patients binge on chocolate when depressed. Could this be an attempt at self-medication — trying instinctively to make for the lowered internal level of PEA? This sounded reasonable, until Dr. Richard Wyatt, a prominent researcher at the National Institute of Mental Health, and a few of his associates tried eating pounds of chocolate, which didn't raise their urine levels of PEA at all and only gave them headaches. Either the stuff is all metabolized in the digestive tract and never reaches the bloodstream, or the wrong people got tested, and we should do the chocolate test on people who say they get a real lift from the stuff. Our hypothesis about the links between love and chocolate has gotten a lot of attention in the press, but it may turn out that people turn to chocolate when they're unhappy for the sugar, the caffeine, or out of habit, and not to get a shot of PEA.

One issue that has to be tackled is why we see more women than men with hysteroid dysphoria, and why many of the men we do see with this problem are gay. While we don't have a complete explanation for this, we do have a few leads.

Men in our society as a group are trained as youngsters to keep a "stiff upper lip" in the face of adversity. Thus it is less acceptable for them to admit to emotional distress or seek psychiatric help. Instead, men tend to abuse alcohol and get into fights as ways of coping with emotional upsets.

Second, men have been enculturated to seek self-esteem, attention, approval, and admiration in the work setting as much as, if not more than, in romance. The opposite has traditionally been true for women. Thus while many of the disappointment-caused depressions we see in women are related to romance, many of those we see in men are related to lack of success on the job. While these sexually stereotyped ways of relating to the world are rapidly changing, they still play major roles in our emotional lives.

As a third possibility, women may be biologically wired from birth to have stronger feelings of attachment, since this same system comes into play when they become mothers. Men and women don't seem to differ much in their attraction mechanisms — hence one sex may get romantically involved as easily as the other. But if the female halves of these interactions have attachment systems that are more readily activated than are those of the male halves, then numerically speaking, women should get hurt more often than men in relationships that don't succeed, and should work harder to keep the peace in relationships that do. No one has actually conducted surveys to see if this is so, but it is a researchable question.

Well then, what do we have to offer to people with hysteroid dysphoria at this time, besides a lot of theories about fluctuating chemicals with impossible names?

The answer, I'm glad to be able to say, is that for most individuals we seem to be able to offer a great deal. In this regard, I've saved the best piece of news for last.

Most people with hysteroid dysphoria are remarkably helped by treatment with a certain type of antidepressants medication called the monoamine oxidase inhibitors (or MAOIs). Moreover, what these drugs seem to do is not just help lift people out of depression, but actually to reduce both the sensitivity to rejection and the craving or need for romantic involvements.

Let me give you a few examples.

Penny W., a twenty-eight-year-old actress, came to see me several years ago. Penny felt that her life had been fine until a year before, when she found out that her husband was having an affair with another woman. Feeling devastated, she left her husband, and, with her four-year-old daughter, moved to Europe. A beautiful woman, she soon began living with a well-known French actor. The new relationship was initially very pleasurable, but within several months she began to feel rejected as he paid less attention to her and at times indicated he did not care so much about her. This would make her quite depressed, but she would medicate herself with diet pills or cocaine, which would quickly restore her mood and energy. Finally tiring of this, she returned to New York and found her own apartment.

Living alone with her daughter for the first time was a new and painful experience for Penny. Although she quickly found work she began to feel desperately lonely, isolated, and depressed.

She gained fifteen pounds, had to take sedatives to calm herself, and often felt suicidal. She had no energy and began isolating herself by refusing to answer the phone. The only things that made her feel better were when a good friend came to stay for a week, or when an audience would be particularly responsive to her dramatic efforts.

Further exploration revealed that Penny had always enjoyed and desperately needed attention from men. In all her previous relationships she'd be elated in the beginning, and when things would begin to go badly she would immediately find someone else. The things she most enjoyed were activities in which she was noticed by others — whether it was acting, going to a party, or being admired in some other way.

Having full responsibility for her daughter was putting a real crimp in this life-style, and this is what led to her long depressive slump. Quite simply, it was hard to party at night or take off for the weekend while trying to work and to look after a young child. The solution of quickly finding a new man to live with when she came back to the U.S. had not worked out. Thus for the first time in her life her usual ways of coping with her "addiction" to attention were blocked, and what resulted were the problems that brought her to see us.

We tried prescribing a MAO inhibitor type of antidepressant for Penny, and it had a dramatic effect. Within several weeks her good mood and high energy returned. She began to see her friends again, put more of herself into her work, started dating several men, and began to enjoy her life.

Also a curious thing happened — Penny found herself enjoying her life even though there was no intense romance going on at the time. She was amazed about this, because for the first time in her life she did not feel driven to get involved with a man. Rather, she was able to enjoy evenings and weekends with her daughter, and began to feel proud of her role as a mother. Men were still interesting and interested in her, but that intense drive for an instant relationship was gone.

Let me add at this point: at the same time Penny started the medication she also started psychotherapy with a female therapist, and also found this helpful. In particular, it helped her begin to identify the patterns in her life, how the same things kept happening over and over again. For example, she was able to see how her craving for a relationship led her to pick her partners very hastily, and that this often led to what were really unsuitable choices for her. Thus for Penny medication and psychotherapy worked hand-in-hand, the former helping to stabilize her mood and self-esteem without constant outside attention and admiration, and the latter helping her to understand her maladaptive behavior patterns. I also think that Penny's seeing a female therapist made it a lot easier for her to profit from the psychotherapy, because there wasn't the same need to be seductive or elicit approval as there would have been with a male therapist.

Another example is Jeffrey G. — a thirty-five-year-old homosexual who came to us complaining of long-standing difficulties in relationships with people as well as chronic feelings of sadness and hopelessness.

In general Jeffrey claimed that he was almost always sad or unhappy, but that periodically he would become much worse when he felt uncared for or unloved. This was often brought on by something unpleasant happening between him and a friend. During his bad periods he would overeat, socially withdraw, not keep up his appearance, feel extremely lethargic, spend a lot of time sleeping, and occasionally use amphetamines. Several years of psychotherapy had not helped very much.

Jeffrey also turned out to be a real "attention junkie." He had been involved in acting since grade school. When involved in a successful performance, he noted his mood and energy were greatly elevated, his appetite much more normal, and he needed only five hours of sleep each night.

What Jeffrey wanted most from treatment was to be able to have a romantic relationship. Although he had been homosexual since high school, he had not had many sexual contacts. In his early twenties he had several romances but each had ended — leaving Jeffrey feeling devastated. More important yet, he never really tried to have another serious relationship since then, feeling that no one else would ever want to be close to him and that he wouldn't want to risk getting hurt like that again.

Once again we tried prescribing a MAO inhibitor for Jeffrey and it has remarkably improved his life. We started the medication several years ago, and the results exceeded all our expectations. Quite simply, Jeffrey has been a much happier man since then. His mood and energy responded quickly, as is usually the case. But what was more impressive was a tremendous change in his self-confidence and self-image. The key, I think, was that he no longer would become depressed when he felt rejected or mistreated. Rather he would simply say, "So-and-so is acting coolly [or cruelly, or whatever the case] — it's his problem, not mine, and I am not going to take it to heart. I'll tell him he hurt my feelings, and we can discuss it." These things just didn't devastate him anymore.

This decreased sensitivity to rejection had profound effects on Jeffrey's life. He developed a wider circle of friends, began to have sexual relationships, and several years ago found a lover — his first in more than ten years.

We've talked about women and gay men, but what about "straight" men? Do some of them have hysteroid dysphoria as well? The answer is yes, although for the reasons I've indicated their difficulties tend to show up more in their work lives than in their love lives.

Jack K. was about forty when I first saw him in consultation. A sharp corporate accountant and fiscal planner, he had been in a several-year slump since being passed over for promotion. Nor was this the first episode; other long depressions had followed earlier work-related disappointments. In addition, Jack was often down for brief periods when he felt that his work was unappreciated or not sufficiently recognized by his superiors.

Jack struck me as a male hysteroid dysphoric, with work substituted for romance. I tried prescribing a MAO inhibitor, which had a highly beneficial effect. We also began to explore his ways of relating to the people with whom he worked. Two years later he was getting along better with colleagues, not nearly as sensitive to the inevitable disappointments that come with working in highly competitive environments, and had actually been promoted several times. While my psychotherapy may have had some effect, Jack becomes depressed and begins to overreact whenever we take him off his medication.

Should we give MAO inhibitors to all highly competitive men who become despondent when things don't break their way? I think not, or we'd have half of corporate America on the stuff. But for people who overreact and slide into deep or prolonged or frequent depressive slumps because of a setback, be it romantic or work related, antidepressant drugs of one type or another may be very helpful. This should not be overlooked by either doctors or patients just because one can point to a precipitating event or personality style that seems to explain the particular episode.

My point in all this is not to convey the impression that the MAO inhibitors are wonder drugs or "magic bullets" or any such thing. Rather, for people with certain types of problems they appear to be extremely helpful. Many of these people have major difficulties in romantic relationships, which is why we are discussing the MAO inhibitors in this book. The fact that some of these patients are helped by drug treatment raises chemical questions about what are usually considered psychological events. As for how long people need to stay on these drugs, it seems to vary. One of my first study patients took a MAO inhibitor for three months four years ago and, with one brief exception, has been fine since. Of the patients described earlier in the book, Jennifer has been on medication for two years and begins to feel more depressed if we lower the dose. Penny stopped after three months and did well, although she also continued psychotherapy for another three months. Jeffrey has been on the drug for two years and is determined to continue. Jack came off for a while, then needed to start again. It would appear that some people can come off after a time, while others may need it indefinitely.

While the examples I have described above are all actual case reports altered only to hide the patients' identities, they were specifically chosen because they did so well and serve to illustrate the points I am trying to make. "Anecdotal" evidence such as this used to be acceptable in psychiatry, but in today's more scientific climate it no longer suffices. And with good reason, because investigators trying to prove something new are naturally biased, tending to focus on the successful cases that support their contentions while paying less attention to those experiences that don't. To really prove our hypotheses about hysteroid dysphoria, rigorous, controlled scientific studies must be conducted.

RECENT RESEARCH FINDINGS

Several such studies have been carried out or are now under way at the Depression Evaluation Service of the New York Psychiatric Institute. This is a treatment and research clinic where patients are evaluated and, if suitable, treated free of charge while participating in various programs.

One study examined the interaction of MAO inhibitors and intensive psychotherapy in women with hysteroid dysphoria. This preliminary study involving sixteen subjects found that ten did extremely well in response to three months of combined MAO inhibitors and psychotherapy. To see whether the drug was necessary, the patients were chosen at random either to continue on the MAO inhibitors or to switch to an identical, inactive placebo, while all continued in psychotherapy for another three months. As is always the case nowadays, all the patients were informed of the study design and the possibility of their being switched to placebo before volunteering for the program. During the second three-month phase both patients and treating staff were unaware of which patients were receiving the active medication and which were getting placebo.

What we found was that at the end of the second three-month period both the active drug and placebo groups had continued to progress or at least maintain the gains made during the first three months of treatment. However, among the six patients switched at the end of three months from active drug to placebo, all had at least temporary mood crashes during the next three-month period. In one case the change was immediate, involving a twenty-seven-year-old model who had come into treatment feeling depressed because of an old boyfriend's lack of romantic interest in her. He visited once while she was still on active medication, and they platonically enjoyed each other's company. During a second visit soon after the switch to placebo, however, the patient began to feel more rejected and began demanding so much attention that her friend cut his visit short and left in anger, at which time she became quite depressed.

Some of the other patients switched to placebo did not nosedive until one to two months after the drug change, so their depressions cannot be attributed to withdrawal from the active drug. Rather, they were no longer "protected" by the medication, and when some painful event transpired, they were liable to react with depression.

Nevertheless, five of six placebo patients ended up emotionally in a good place at the end of the sixth month, which is why the placebo group as a whole did as well as the group that stayed on active medication. This would suggest that the combination of three months' medication and six months' psychotherapy was an effective treatment program for these women.

Does this mean that everyone with hysteroid dysphoria should be out on medication for at least three months? To answer this scientifically would require a study involving a large number of patients and a different research plan. Ideally, one would want to compare four different situations: psychotherapy without medication, medication without psychotherapy, psychotherapy plus medication, and no active treatment. From a technical point of view this would be a very difficult study because just as we give placebo pills to disguise from patient and doctor alike who is getting real medicine, so placebo psychotherapy must be devised as a comparison for the real thing. Problems like this make good long-term psychotherapy research extremely difficult, expensive, and time-consuming to do. In the meantime, I try to inform my patients of the potential benefits and risks of taking MAO inhibitors (or other types of antidepressant) and let them decide for themselves.

THE MAO INHIBITORS

In talking about the MAO inhibitors we are not talking about a new class of antidepressant drug, but rather, a new use for an old drug. As a matter of fact, the first MAO inhibitor was synthesized in the early 1950s. Found useful as an antidepressant, it was abandoned because it sometimes caused liver damage. But other MAO inhibitor type drugs were developed and found helpful for what have been called "atypical depressions" — those in which people were also anxious and very emotional, often felt worse in the evenings, had trouble falling asleep or overslept, often overate, and could still be cheered up. These symptoms differ from those in more classic or "typical" depression where a person lost his appetite, could fall asleep but not stay asleep, and couldn't be cheered up at all. However, the MAO inhibitors seemed to have this one terrible problem — people on them would get unexplained and sudden elevations of blood pressure that caused terrible headaches and sometimes even strokes or death. So while the drugs appeared helpful, they almost disappeared from use by the late 1960s.

What has brought them back was the discovery that the high blood pressure reactions were due to people eating or drinking certain foods or taking certain medications while on a MAO inhibitor. What the MAO inhibitors do is to inhibit or block a certain enzyme in the body — monoamine oxidase. This enzyme is responsible for breaking down certain chemicals that help the brain transmit messages. Inhibiting the enzyme blocks this removal or metabolic process, and the level of these neurotransmitter substances go way up. In general, their increase is not a problem, since they are stored in "pools" in nerve endings and released to have their effects only a little at a time. But certain substances found in some foods cause these storage pools of neurotransmitters to empty and flood the brain. This produces overwhelming activity of the sympathetic nervous system (the "fight-flight" system), causing among other things rapid elevation of blood pressure.

People taking MAO inhibitors are at risk for a high blood

pressure reaction for two reasons. One is that their neurotransmitters' storage pools are very full, so that anything that suddenly empties them has a massive effect. Second, they have lost their protection against tyramine, a substance found in many foods that can cause these storage pools to suddenly dump their contents, but which in people not taking MAO inhibitors is broken down by the enzyme monoamine oxidase lining our digestive tract.

The risks of high blood pressure reaction are minimized by placing all people taking MAO inhibitor type drugs on special diets, which prohibit foods and beverages high in tyramine. Among the banned substances are aged cheeses, red wines, beer, fava beans, pickled herring and pickled lox, and meat, fish or poultry that it not fresh, freshly frozen, or freshly canned (e.g., game meats, smoked meats, aged steaks). Stimulant and decongestant medications are also prohibited. People do well if they follow the diet and most patients have no trouble. Minor lapses produce only a bad headache that lasts for thirty minutes or so. However, anyone considering taking a MAO inhibitor type drug must be committed to following the dietary restrictions, since a major lapse could cause a stroke or even a fatal brain hemorrhage, and we don't give these medications to people who can't or won't be reliable in this regard. Good news for the future, however, is that we are actively working on newer MAO inhibitor type drugs that will have the same beneficial effects without the dangerous or troubling side effects.

I think these findings are significant in a number of ways. For one, the MAOIs alone or in combination with psychotherapy appear to be of significant benefit to a number of individuals with the kind of problem we call hysteroid dysphoria. This in turn suggests a whole new "chemical" way of thinking about people with extreme needs for romantic (or other types) of attention and involvement, as well as with extreme sensitivity to rejection. Although how the MAO inhibitors help dampen or correct these vulnerabilities is not known at this time, we speculate that the drugs prevent the dramatic fluctuation of some internal chemical substance, perhaps phenylethylamine or perhaps something else — something that causes energy, mood, sense of well-being, and so on to shift in response to social interactions. In particular the drugs seem to prevent or diminish the lows, perhaps by preventing the level of this chemical from falling below a certain point. It would not prevent the body from making more of the substance in socially rewarding situations, however, and people with hysteroid dysphoria taking MAO inhibitors don't report any diminution in their ability to feel joy, excitement, or love. If anything, the drug dosage may need to be lowered at times to prevent a person from feeling too high.

We started out this chapter by saying that everyone is saddened and hurt to some degree by the breakup of a romance. We have no effective way to prevent that. But heartbreaks come in many shapes and sizes; what I've tried to provide is a brief overview of the common patterns and ways to think about, cope with, or get help for each of them. For while a certain degree of suffering is part and parcel of romantic involvement, much that occurs may be unnecessary and avoidable. Toward this end, thinking about romantic ups and downs in biological as well as psychological terms is extremely helpful.

CHAPTER 7

On Love

There has been a tendency in some quarters of late to devalue romantic love, with the critics divided into several schools of thought. One group sees a preoccupation with romance as a particular cultural craze or addiction that has afflicted Western civilization for the past several generations and has yet to peak — an emotional "hula hoop" that won't go away. A second type of critique grew out of the Freudian tendency (but does not include many modern Freudians) to look at romantic love either as a substitute for our true instinctual desires or as the workings of neurotic mechanisms; this can be called the "Romance as Neurosis" argument. The third critique, a more recent trend, sees romantic love as a male sexist invention designed to keep women from achieving their potential and, in the process, competing for what are now male prerogatives. All these viewpoints have some validity but, on balance, miss the forest for the trees. That is, they focus on one or another problematic aspect of romantic love, and then treat that as a total explanation.

While it is true that our society is focused on romance to an extent not seen in other cultures or in other periods of history, it is not true that twentieth-century America or even Western civilization holds the patent on romantic love. As a number of anthropologists and historians have shown, romance is found in many Polynesian and African cultures and in other periods in history such as eleventh-century and eighteenth-century Europe. What is unique about our twentieth-century Western world is that we are the first society to hold the widespread belief that if you don't love someone, you shouldn't marry him/her.

Throughout human history, marriage has not been seen as a vehicle for expressing love and devotion, but rather as a way of cementing alliances between families or creating viable economic units who can support themselves and raise families. Many human societies have elaborate rules about whom one may or may not marry. Some involve incest taboos, which may serve the function of preventing marriage between close blood relatives, with its liability for inherited diseases. But principally who one may marry has been controlled by families as a way of forming or cementing alliances, enhancing (or at least not diminishing) the family's wealth or status, and ensuring that its traditions continue on into succeeding generations.

This is seen across many cultures, particularly among the upper classes, who have the most to lose if their children marry incorrectly. Not surprisingly, upper-class families also have the greatest means to control who their children do marry, by agreeing or refusing to pass on their wealth, titles, and privileges. Poorer families, again across the globe and throughout history, have been more concerned with ensuring that their children can feed, clothe, and shelter themselves.

The problem with romantic love is that it represents a threat to family control of whom children marry. It's no good telling Romeo or Juliet that more suitable partners will be found for them when they've already fallen in love with each other. If children are totally free to choose their marital partners on the basis of love, then all community and familial concerns about maintaining alliances, merging wealth, and passing on traditions (religious or otherwise) are at risk. So love has had to be controlled.

How this has been done is a testimony to human ingenuity. Many ways of channeling romantic love have sprung up around the globe, and all societies, including our own, use one or more of them. Some, like traditional Hindu society, rely on child betrothal; this is one of the more drastic forms, and essentially involves arranging the marriage before the kids know what's up. Any love that grows up between marital partners takes place after the fact.

Other cultures rely on other control techniques, such as rigidly designating the categories of people one may marry, setting a price that either the bride's family or the groom's family must pay, or in some other way limiting the pool of eligible partners. We may like to think of ourselves as free to marry anyone we choose, and theoretically we may be. But in actuality we are trained to regard a suitable partner as possessing certain qualities and also raised in neighborhoods and sent to schools populated by the kind of people our parents want us to marry. Thus both our thoughts about suitable partners and our opportunities for meeting them are strongly influenced by our family upbringing. Most Americans marry people of the same racial, religious, and social background. Even in romance-crazy America love is channeled in certain directions.

What makes our society unique is that people for the most part feel they have to love someone to marry him or her and, increasingly, to stay with that spouse. In other societies, as far as romance was concerned, what one tended to see was very, very few people marrying for love, more falling in love with their husbands or wives following the marriage, and still others falling in love with someone else after being married for a while.

This last possibility, which could be a threat to marital

stability, has also been handled in a variety of ways. One way, which grew up with the troubadours, knights and ladies in eleventh-century Europe, was to romanticize extramarital attachments but keep them free of sex — this is what we know as the doctrine of courtly love. Seven hundred years later, in eighteenth-century Europe, the upper classes blatantly indulged in extramarital affairs that did include sex. These were tolerated by the society as a whole and did not threaten marital stability. This practice is also seen in many primitive societies, such as the Turu tribe of Tanzania, where, as long as certain face-saving gestures are made, extramarital romantic liaisons are considered routine.

Many Americans tend to follow the same pattern, and many marriages no longer based on love are sustained by extramarital affairs. However, both as individuals and as a society we are much less open in acknowledging this, partly because of our Judeo-Christian heritage, and partly because we believe that love and marriage should be linked. Interestingly, marriages that continue for convenience seem to be more common among upperclass Americans and Europeans, who again have more to lose from marital disruption.

That marriage should be based on love is also straining marriage as an institution. People expecting to be in love before they marry is not so new — we have, as I've said, ways to channel this. Also, it fits with our economic needs for mobile nuclear families in place of more rooted and extended kinship groupings (who would traditionally arrange the marriages and offer emotional support that we now seek and expect from our spouses). But what is new is the notion that people should not remain together unless they continue to love each other, a belief that was much less widely held a generation ago and which is responsible for our soaring rates of divorce and resultant single-parent households.

While this confronts us with a new set of psychological and social problems, it does not provide evidence that we are in the throes of some romantic craze or set of feelings unknown in other cultures or periods of history. Rather, what we are experiencing is the logical outgrowth of a society in which traditional structures such as church and extended family play an ever-diminishing role in our lives, where individual freedom (political, economic and social) has steadily increased, where the status and respect accorded to women has steadily advanced, and where relative freedom from economic want and growing abilities to support oneself for large segments of our society has left us freer to pursue emotional fulfillment through romantic love. That the media now use this to sell everything from toothpaste to automobiles may influence our notions about romantic love, but are not their source. That many people turn to romance to fill in the loneliness or isolation brought on by loss of stable community and extended family ties is also true, but again, this is not the cause of our concern with romantic love. In fact, we can also turn this argument around and say that romantic impulses (as well as economic ones) are what have led modern Americans to shake loose from those community and family ties. Rather, romantic love appears to be a universal human potential or capacity that has been suppressed in some societies, channeled to one degree or another in all, and seems to be flowering of late in ours.

Freud's view of the human condition was a rather gloomy one. He believed that our basic instincts, which he saw as sex and aggression, were incompatible with life in a civilized society and could be only partially satisfied. Since full sexual satisfaction was impossible, sexual energy built up and was transformed into romantic feelings. While modern psychoanalysts have largely abandoned this hydraulic model of the way people act, vestiges of it remain, especially when romantic love is being considered. One problem with this is that Freud's original work was done in an era of Victorian sexual restrictiveness. Our own era, with its sexual freedom and romantic preoccupations, doesn't fit the "love as rechanneled sex" model very well.

Other psychological theorists have viewed our adult search for an intimate partner as an attempt to recreate the bliss of our early mother–child bond. As adults, we are said still to yearn for this lost early state and to try to find (actually "refind") someone to have it with. For those romantics who feel their early attachment experience was not so hot, the theory has it that they are seeking someone who will help make up for their earlier deprivation.

Romantic love seems to be much more than this. One problem may be that many mental health practitioners spend much of their time dealing with troubled people, and hence may get a biased view of the human condition. My point is that if you want to understand something like romantic love, it's important to study people who function happily as well as those with emotional problems.

Romantic love is a normal and healthy emotional state. Moreover, it is a crucial emotional step for young adults *once* they have more or less found their individual identities. (One's own identity comes first because one needs a firm sense of oneself before one can function successfully in a romance.) As a former teacher of mine, Dr. Otto Kernberg, has written, sexual intimacy in a loving relationship is one of the few opportunities we have to cross the barriers that normally separate us from others. At its best, romance stretches us as human beings, allowing us to feel genuine tenderness and empathy and to share goals and values. In this vein, falling in love can help people grow and mature, and may even overcome the remains of certain psychological problems.

From the perspective of modern psychology, then, our approach to romance can be healthy or unhealthy. If you are relying on others to meet too many emotional needs (such as needing another person in order for you to feel good or worthwhile), expecting the relationship to be a continual honeymoon, not picking and choosing whom you get involved with, or not having a good sense of your own identity apart from your partner, then you will be headed for rough sailing when you get involved romantically.

The flip side of this is that for people who have taken the time to develop individual identities and interests, are thoughtful about their choice of partners, can say "I've had enough" if that is what needs to be said, and can focus on someone else's needs beside their own, romantic love can be a source of joy and enrichment.

The only problem with this modern psychological view of romantic love is that *it needs to be supplemented by an understanding of the biological side of our romantic capacities.* The basic issue is that a healthy approach to romance may be easy to describe but much more difficult actually to achieve, and we have to struggle with biological as well as psychological vulnerabilities when it comes to romance. We need to marshal all our knowledge, be it cultural, psychological, or biological, to handle the tumult of falling in love, the problems of keeping a relationship alive and vital, and the terrible emotional pain we feel when things are not going well or fall apart completely.

Is romantic love a male "sexist" invention or a ploy women have had to resort to because they lack power in our society? Many strong feminists have, in recent times, come to view romance as a psychological trap whose aim is to prevent women from achieving their full individual potential. While my own views obviously reflect a male perspective, I see no basic contradiction between equality for women and romantic love. In fact, it is precisely the growth of the rights, freedom, and respect accorded to women over the course of our history that has made romantic love such a widespread possibility in our times. What we are witnessing now, I believe, is some dislocation of traditional male–female roles, which is creating new strains on relationships. For if women are to pursue the career-oriented lives that many men now lead, something has to give, or who is going to look after the home and raise the children?

One alternative for some people is not to have children. Another, if you're wealthy, is to hire help. (One married woman physician, often torn between staying late at work and leaving to pick up her son after school, exclaimed in a moment of frustration: "I need a wife.") A third possibility is for men and women to share tasks more equally, which ultimately may necessitate some modification of the way men expect, and are expected, to pursue their careers. What I hope to see develop over the next generation or two is a greater range of comfortable options, all the way from relationships that share tasks rather equally, to couples where either the man or the woman stays home and where efforts are rather strongly divided along those lines.

Romantic love is a basic human potential or capacity, something that we have been wired for for eons but only free to pursue on a broad scale in recent generations. Love involves a complex set of biological processes that we are only beginning to understand, but which appear to involve distinct attraction and attachment mechanisms. The allure of romance is, in part, due to the power with which these processes, once set in motion, can affect us.

After some reflection, an appropriate image from Greek mythology came to mind. It is the story of Phaethon and the sun chariot. Phaethon was son of Helios, the sun god. Every morning Helios emerged from the east in a golden chariot to ride across the skies and light up the world. The chariot itself was drawn by eight dazzling white winged horses whose nostrils breathed forth flame. One day some of Phaethon's companions challenged his claim to divine parentage, and he went to Helios to seek proof. Helios, wanting to reassure his son, offered to grant the boy any wish in his power. Phaethon requested to be allowed to drive the sun chariot. Horrified by this request because he knew the boy could not control the powerful horses, Helios begged Phaethon to alter his wish, but the boy was adamant. With a heavy heart Helios let Phaethon take his place the following morning. Once under way the horses quickly sensed Phaethon's light and inexperienced hand on the reins, and began to run amok across the skies, sometimes going too low and scorching the earth, sometimes too high and freezing whole regions. Finally in desperation Zeus was forced to hurl a thunderbolt at the careening chariot, destroying Phaethon.

The image I have of romantic love is that of the sun chariot drawn by eight fire-breathing winged steeds. The moving chariot can overwhelm some drivers because things can quickly get out of control. Romantic involvements are due to, and further cause, the unleashing of a powerful set of biological and psychological forces within each of us. Being prepared for the journey, in this case a romantic involvement, demands both a certain degree of psychological maturity and biological stability, as well as some basic understanding of relationships.

As we are just beginning to comprehend, many aspects of human behavior are shaped by a complex interaction of our biological capacities, psychological development, and cultural heritage. Nowhere is this more true than when it comes to romantic love. Understanding and utilizing this knowledge will, I hope, make for a smoother ride.

Afterword

In the final analysis, how we feel about our lives depends on how much joy, excitement and pleasure we experience and how much anxiety we suffer. While the actual things that give pleasure or stir up anxiety may differ for each of us, ultimately these feelings are shaped by the workings of our brain pleasure and anxiety centers and circuits.

Nowhere is this more true than for our deepest romantic feelings. The thrilling high of falling in love, the desperate anguish of a broken romance, the warm security of an established relationship, the unsettled feeling of a serious spat — all these are due to chemical changes in the brain stirred up by interactions with someone who has become important to us. Or perhaps it would be more accurate to say someone has become important to us because of the ways he or she affects our brain chemistry.

But how does this help us understand why so many people have so much trouble with their love lives? Why is it so hard to meet the right person? Why are marriages coming apart at an ever-increasing rate?

There are three reasons. Most of us have been badly programmed culturally when it comes to romance. Specifically, we live in a society that tells us a lot about falling in love and virtually nothing about staying in love. Falling in love is easy — you meet someone who looks and acts in a certain way, and soon your pleasure centers are firing like crazy, just like in the movies. But the movies never tell you what to do next, except to head for the nearest sunset. But how to make a go of things in the years that follow the honeymoon is a much thornier issue, because our biological wiring seems to be such that, over time, our feelings of excitement for a long-term partner level off. What holds relationships together instead are our feelings of attachment, which have more to do with security than excitement.

But to keep a relationship vital you still need some romance, some passion, some thrills. You need to keep things moving. This means building bridges, breaking down barriers, growing as individuals, learning how to fight and make up, and realizing that other attractions, no matter how enticing, are usually better left alone. If you want a situation where you and your long-term partner can still get very excited about each other (as most of us do in this day and age), you have to work at it, because in some ways you are bucking a biological tide.

In addition to our unrealistic cultural programming, some of us have been individually programmed (usually in childhood) in ways that make life, and love, unnecessarily difficult. There are many possibilities, such as being taught to feel guilty if you are having too much fun or that you're worthwhile only if you're smarter or better looking than everyone else. This is the kind of stuff to see a psychotherapist about.

The third possibility is that your emotional wiring itself is faulty. For many years we've recognized that some people have pleasure centers that shut down or rev up periodically for no apparent reason, producing the emotional states we call depression and mania. But it now looks as though many people with romantic difficulties have brain stimulant systems that overreact, making them so giddy they can't think clearly when they meet an attractive person, or cause them to crash into depression in the face of a romantic disappointment. Others, with different biological instabilities, seem to suffer chronic feelings of depression or addictions to romance, drugs, or anything else that offers even a temporary lift.

Anxiety circuits can also function improperly. Here our knowledge is even more preliminary. Yet it is beginning to appear that a variety of problems, such as severe panic attacks, exaggerated fears about loneliness or separation, and perhaps even the tendency to worry excessively, are all rooted in the faulty functioning of certain brain alarm systems.

These emotional wiring problems are what modern psychiatric research is more and more gearing up to tackle. A number of major medical centers have recently set up specialized depression units, which, although they aren't advertised this way, are really in the business of evaluating and treating disturbances of our brain pleasure systems. While the basic way we size up these problems is still the clinical interview, some blood and urine tests look very promising, not to mention more science-fiction-sounding things (like the PET Scanner) that are farther down the road. Moreover, our ability to help people with certain biologically based romantic difficulties is growing all the time, especially with the use of medications.

For reasons that are not entirely clear, research in the anxiety field has lagged behind. Because of this, we have recently set up an Anxiety Disorders Clinic at the Columbia Presbyterian Medical Center and New York State Psychiatric Institute, to study and treat people with many different kinds of anxiety problems. Studies are already in progress for people with panic attacks, phobias, intrusive thoughts (obsessions), or uncontrollable urges to repeat things over and over again (compulsions).

But other patterns, such as excessive fears of being abandoned or being assertive, as well as chronic worry, will also be studied. When you get right down to it, brain chemistry is what separates the worriers from the warriors among us.

Although I'm not about to start one, I can also conjure up a futuristic dating service, where potential partners would be matched for biological compatability. Actually people now often do this intuitively, since it makes sense that romantic partners should be alike in how much they crave novelty or excitement, what kinds of things they get turned on by, and in their capacities for peak experiences. While figuring out exactly which chemicals or brain pathways we might actually measure is still a thing of the future, we do have some leads.

Along with my scientific enthusiasm, however, come several concerns. The first, which I raised earlier, is that knowing how our brains work will somehow ruin the poetry of our lives. This really doesn't worry me very much, since knowing how cars or digestion works doesn't seem to diminish the pleasure we get from driving or eating. Ultimately, we are not striving to remove life's poetry, but to enable more people to experience it.

Something that does concern me, however, is whether a clearer chemical understanding of our emotions could someday be used to harm rather than help mankind. As we have seen, the basic impetus for this work has been scientific and medical — to understand ourselves better and to help people with emotional difficulties. But as our understanding of the biochemistry of romance and other emotional states increases, one has to think of all the ways this information could be used. Could this knowledge of brain chemistry be used, for example, to manipulate or control our thought processes, or to change or blunt our emotional loyalties or ties?

This has to be considered in light of some of the uses to which we have put other scientific breakthroughs. Like it or not, scientific understanding of our emotional chemistry is presently growing by leaps and bounds, and this will only be more true in the years ahead. What must therefore occur simultaneously is the continual insistence on ethical investigations and humanitarian application of this knowledge. While scientists themselves must always feel responsible for the conduct and impact of their work, the ethics committees that now oversee all research activities at our major medical centers are also helpful to this process. A third vital component is an informed public.

My ultimate hope is that anything that gives us a clearer understanding of how human beings think and feel will, in the long run, help us make our world a better place. Most of what creates so much difficulty for us as a species, such as our proneness to rage, violence, distrust and acquisitiveness, seems to be based in our brain chemical systems. Thus, the more we know about how we work emotionally, the better able we may be to deal with the darker sides of our human personality. In the end, however, it is our capacity to love that holds out the most hope for mankind.

Suggestions For Further Reading

What follows are brief descriptions of some of the books and articles that I would recommend to interested readers. It is not meant to be an all-inclusive list of references.

BOOKS

Poetry and Prose of William Blake
Ed. Geoffrey Keynes, The Nonesuch Library, London, 1961.
A poet's understanding of human emotions and mental functioning that scientists are only beginning to reach.

Attachment and Loss
By John Bowlby, Vol. I *Attachment*, Basic Books, New York, 1969.
A detailed presentation of how our capacity for attachment and reactions to separation may be "wired in" from birth.

Goodman and Gilman's The Pharmacological Basis of Therapeutics
6th ed., ed. A. G. Gilman, L. S. Goodman, and A. Gilman. Macmillan, New York, 1980.
A standard reference work for medical people, the chapter on drug addiction and drug abuse (chapter 23) is well written and informative.

Toward a Psychology of Being
By Abraham Maslow, D. Van Nostrand Co., Princeton, 1962.
A humanistic psychologist's description of our capacity for peak experience.

Sex and Temperament in Three Primitive Societies
By Margaret Mead, Morrow, New York, 1963.
A nice introduction into the romantic practices of non-Western societies.

There is no single source for all non-Western cultures, and interested readers will have to dig through the anthropology literature.

Love Sex and Marriage Through the Ages
By Bernard L. Murstein, Springer, New York, 1974.
The most comprehensive history of romantic love available.

The Passions
By S. G. Milligan, Darling, London, 1848.
A surprisingly modern understanding of the biological basis of emotional disorders by a nineteenth-century psychiatrist.

The Brain: The Last Frontier
By Richard M. Restek, Warner, New York, 1979.
A comprehensive and well-written presentation of what scientists are learning about the human brain.

A New Look at Love
By Elaine Walster and G. William Walster, Addison Wesley, Reading, Mass., 1978.
A more psychological approach to romantic love by two prominent researchers.

Mind Mood and Medicine: A Guide to the New Biopsychiatry
By Paul H. Wender and Donald F. Klein, Farrar Straus Giroux, New York, 1981.
A book for the general reader by two distinguished psychiatrists, who explore the exciting new findings in the cause and treatment of mental illness that are currently revolutionizing mental health practice.

ARTICLES

"The Theoretical Importance of Love"
By William J. Goode, in Winch et al. (ed.), *Selected Studies in Marriage and the Family.* Holt, Rinehart and Winston, 1962 (pp. 455–476).
A perceptive discussion of what romantic love represents, and how it has been handled by various human societies.

"Intracranial Self-stimulation Thresholds"
By Conan Kornetsky, Ralph Esposito, Stafford McLean and Joseph Jacobson, in *Archives of General Psychiatry*, Vol. 36, March 1979 (pp. 285–292).
By linking drug abuse in humans to the effects of the same drugs on self-stimulation patterns in animals, the central role of human "pleasure centers" is highlighted.

"The Styles of Loving"
By John Alan Lee, in *Psychology Today*, October 1974 (pp. 44–51).
A thoughtful description of the different patterns of loving.

"New and Old Evidence for the Involvement of a Brain Norepinephrine System in Anxiety"
By D. E. Redmond, Jr., in *Phenomenology and Treatment of Anxiety*, ed. by W. E. Fann, I. Karacan, A. D. Pokorny, and R. L. Williams. Spectrum Publications, 1979.
An argument for the locus ceruleus area of the brain being the center of our alarm circuitry.

"The Background of Safety"
By Joseph Sandler, in *International Journal of Psychoanalysis*, Vol. 41, 1960 (pp. 352–356).
Views feeling states, particularly the pursuit of a feeling of safety, as the prime mover of both our behavior and our fantasy life.

"Cognitive, Social and Physiological Determinants of Emotional States"
By Stanley Schacter and Jerome Singer, in *Psychological Review*, Vol. 69, No. 5, September 1962 (pp. 379–399).
A paper that had a great impact when published. It suggested that emotional states were more influenced by cognitive cues than biological changes.

"Dimensions of Sensation Seeking"
By Marvin Zuckerman, in *Journal of Consulting and Clinical Psychology*, Vol. 36, 1971 (pp. 45–52).
One of the early papers describing the different patterns of sensation seeking.

"Sensation Seeking and Its Biological Correlates"
By Marvin Zuckerman, Monte S. Buchsbaum, and Dennis L. Murphy, in *Psychological Bulletin*, Vol. 88, No. 1 (pp. 187–214).
A review of the trait of sensation seeking and the biological phenomena associated with it.

Index

abandonment, fear of, 64, 114, 143, 202; *see also* separation anxiety

acupuncture, 58

adaptation. *See* tolerance

addiction: to attachment, 136–137; to attention, 5, 102, 167, 170; to attraction, 136, 137; to drugs, 60–62, 65–67, 69, 135, 136, 201

adrenaline, 25–27, 34

alarm center of brain: and panic, 112–113, 114; and romantic attraction, 93

alcohol, 59, 62, 68, 72–73, 156

amphetamine, 18, 66, 80, 123; effect on brain/blood/nervous system, 37, 38, 41–42, 58, 68–69, 70, 176, 177; effect of compared to romantic attraction, 92, 96–97, 99, 106; reverse tolerance, 62

anger, physical sensations accompanying, 22–24

antianxiety agents, 68, 70, 72; antidepressants as, 156, 157; naturally occurring in body, 37

antidepressants: and alarm center, 112–113; development of, 21; effect on anxiety, panic, 147, 156–157; effect on personal relations, 8, 9, 74–75, 86–87; experiment with, 155–156; and hypomania, 98; MAO inhibitors as, 82; and melancholia, 164; normalize brain chemistry, 12–13, 58, 68, 74–76; and postsynaptic receptors, 36

antimanic agents, 12–13

Antiochus, 151

antipanic agents, 68, 112

antipsychotic agents, 13, 21, 155

anxiety, 10–11, 37, 42, 43, 48, 68, 70, 72, 156, 157, 201–202; *see also* panic attacks; separation anxiety

Anxiety Disorders Clinic, 201

"applause" syndrome, 13, 102, 176

attachment, in love, 90, 95, 103, 106, 117, 119, 120; excessive need for, 136–137, 144–145

attention, excess need for, 5, 102, 167, 170; *see also* hysteroid dysphoria

attraction, in romantic love, 90, 91–96, 101–102, 117, 123–127, 141–142; biochemical factors involved in, 96–101, 119–120, 152; excessive need for, 136, 137; lithium and, 158–159

average evoked response (AER) augmenters, 77, 119

average evoked response (AER) reducers, 77

avoidance: as motivation, 42, 44; as post-romantic response, 165–166, 171–172, 182

axons, 32

barbiturates, 21, 68, 72

Berscheid, Ellen, 124

biochemistry. *See* brain; brain chemistry

Bion, Wildred, 166

Blake, William, 17

blood: cortisol and stress/depression, 19–20; metabolization of drugs, 59, 60, 68–69, 176, 177

B-love (Maslow), 120–121

boredom, 22, 42, 43, 76–77; avoidance of in love relationship, 137–142

Bowlby, John, 108–109

brain, human: alarm center, 93, 112–113, 114; central nervous system, 32–33; cortico-limbic programs, 126, 133; described, 38–41; displeasure center, 41, 43–44, 45–48, 93; "emotional wiring" of, 37–38, 40–49; experiments on, 51–52; naturally occurring drugs in, 36–38, 71–72 (*see also* endorphin; enkephalin); neurotransmitters, 34–35, 58, 81, 155, 158, 187–188; and PET Scanner, 54–55; pleasure center, 41–44, 46–48, 51–55, 66, 74–75, 79–80, 93, 99, 146, 163, 200–201; postsynaptic receptors, 35–37, 57–58, 60; synapses, 34–35; *see also* brain chemistry

brain chemistry: antidepressants as normalizers of, 12–13, 58, 68, 74–75; and depression, 19–20, 34, 35, 51, 74–75, 82, 98–100; drug and nondrug effects compared, 55–57, 63–68, 91–92, 96–97, 99, 106; effect of drugs on, 16–17, 57–65, 68–69, 70, 176, 177; endorphins, enkephalins, and naturally occurring drugs in, 36–38, 56–58, 71–72, 106, 107, 113, 134, 148, 157; and emotions, 38, 40–44, 46–49, 90, 96 (*see also* and romantic love, *below*); and hysteroid dysphoria, 176–177; and mania, 80–81, 98; MAO and MAO inhibitors,

209

Index

brain chemistry (*Cont.*)
81–83, 119, 120, 157, 179–180, 182–187; phenylethylamine, 37–38, 81, 99–101, 177–178; pleasure and displeasure centers, 41–44, 45–48, 51–55, 74–75, 79–80, 93, 146, 163, 200–201; ramp effect, 130, 131; rebound effect, 134; and romantic love, 4–5, 7–8, 31, 48–49, 89–93, 96–101, 119–120, 152; and separation anxiety, 106–107; withdrawal effect, 63–64, 132–133, 146; *see also* brain, human

caffeine, 58
CAT Scanner, 54
central nervous system, 32
cerebellum, 77
cerebral hemisphere, 40
challenge, as trigger to romantic attraction, 123–124
children and childhood: dependence, 136; fears, 110, 111; emotional patterns set in, 45–47, 75, 145, 162–163, 200; separation anxiety in, 104–105, 108–111, 143
chocolate, 100, 169, 177–178
clinging, 110, 111, 143, 144–145
cocaine, 18, 20–21, 58, 61, 62, 65, 67, 68, 69, 70, 80, 123
Coca-Cola, 21
codeine, 70
compulsions, 201
compulsive socializing, 11, 143
consciousness-raising, 146, 147
Coolidge, Calvin, 129
cortico-limbic programs, 126, 133
cortex, 40
cortisol, 19–20
cyclotron, 54; 55

delirium tremens, 62
delusions, 18, 35; delusional loving, 153–155
dendrites, 32
deoxyglucose, 54
dependence, drug: contrasted to addiction, 135–136; physical, 62–63; psychological, 63
dependency, in romance, 3–4, 110–112; excessive, 136–137, 143, 144–145
depression: amphetamine and, 68, 69; antidepressants' effect on, 8, 9, 12–13, 21, 58, 68, 74–76, 86–87, 164; biochemistry of, 19–20, 34, 35, 51, 74–75, 82, 98–100; hysteroid dysphoria and, 166–167, 168, 169, 170; melancholia, 48, 74–

75, 163–164; in romance breakup, 5, 162–164, 201
Depression Evaluation Center, New York State Psychiatric Institute, 163, 184–185
"Diagnosis of Love Sickness . . . , The" (Mesulam and Perry), 150–151
disappointment reaction, 164
disinhibition, 73, 76
displeasure centers of brain, 41, 43–44, 45–48, 93
D-love (Maslow), 120, 121
Doors of Perception (Huxley), 16–17
dopamine, 37, 58, 81, 98; and psychosis, 35, 155
drugs: addiction to, 60–62, 65–67, 69, 135, 136, 201; dependence, 62–63, 135–136; effects of, compared to nondrug experience, 55–58, 63–70, 92, 96–97, 99, 106; metabolization of, 59, 60, 68–69, 176, 177; naturally occurring in body, 36–38, 71–72 (*see also* endorphin; enkephalin); placebos, 56–57, 185–186; and pleasure center, 55, 66; and postsynaptic receptors, 57–58, 60, 71–72; psychoactive, 57–59, 60–61 (*see also* specific drugs); ramp effect, 130; rebound effect, 64–65; tolerance to, 60–61; withdrawal, 59, 60, 62–63, 132, 135; *see also* specific drugs, drug categories

electroencephalograph (EEG), 52, 77
electromyograph (EMG), 53
emotions, 22–30, 76, 164; and brain chemistry, 38, 40–44, 46–49, 90, 96; effect of childhood on, 45–47, 75, 145, 162–163, 200; effect of drugs compared to, 55–57, 63–68, 91–92, 96–97, 99, 106; and memory/prior experience, 44–48; tolerance, 61–62; *see also* anxiety; depression; love; romantic love; separation anxiety
endorphin, 37, 56, 57, 58, 106, 107, 113, 133–134, 157, 148
enkephalin, 37, 56, 58
enzymes, 81, 187; *see also* specific enzymes
Erasistratos, 151
experience, 23, 27–28, 33, 38, 42–48, 61, 162–163
"experience seekers," 76, 77

facial expression, 27, 52–53, 55
fantasy: in romantic attraction, 122–123, 129; and delusional love, 153–155

Index

monoamine oxidase inhibitors. *See* MAO (monoamine oxidase) inhibitors
Money, John, 141
morphine, 37, 58, 68, 70, 157
motivation: for behavior, 42–43; for attraction, 25–30, 122–123
mystical experiences, 16–17, 67, 115

naloxone, 56, 57
narcotics, 13, 57–59, 65–66, 68, 70–71; naturally occurring in body, 36–38, 71–72 (*see also* endorphin; enkephalin)
narcotics blockers, 70
National Institute of Mental Health (NIMH), 82–83, 100, 177
nervous system, 32–33; neurotransmitters, 34–35, 58, 81, 155, 158, 187–188; postsynaptic receptors, 35–37, 57–58, 60; synapses, 34–35
neurochemistry. *See* brain chemistry
neurotransmitters, 34–35, 58, 81, 155, 158, 187–188
New York Times, 100
norepinephrine, 34, 35, 37, 58, 81, 98
novelty: as factor in romance, 128–131; in maintaining love relationship, 137–142

obsessions, 201; in delusional love, 153–155; in psychosis, 18, 153, 155
"old boyfriend/girlfriend" syndrome, 131–132
Olds, James, 41
opium, 20, 68, 70; postsynaptic receptors for, 36, 37, 58
overparenting, 110–111, 114

pain, physical, 58; transmission by nervous system, 32–33
painkillers, 37, 56–57, 70
panic attacks, 10–11, 106, 111–112, 147, 156–157, 201; *see also* anxiety
Panksepp, Jack, 106
paranoia, 68, 153
PEA. *See* phenylethylamine
perception: effect of psychedelics on, 15–18; of past experience, as factor in emotions, 23, 27–28, 33, 38, 42–48, 61, 162–163
peripheral nervous system, 32
Perry, Jon, 151
PET (positive emission tomography) Scanner, 54–55, 201
Phaethon, 197–198
phantom lover syndrome, 153

phenylethylamine (PEA), 37–38, 81, 99–101, 177–178
phobias, 201
physical dependence, on drugs, 62–63; contrasted to addiction, 135–136
physiological reactions: as part of emotion, 22–24, 28–30; and sympathetic nervous system, 24–25; transmitted by nervous system, 32–33
placebos, 56–57, 185–186
pleasure center of brain, 41–44, 46–48, 66, 93, 99; efforts to measure, 51–55; malfunctions and shutdown of, 51, 74–75, 79–80, 146, 163, 200–201
positive emission tomography. *See* PET Scanner
postsynaptic receptors, 35–37, 57–58, 60, 71–72
problem-solving with partner, 140–141
Proxmire, William, 84
psilocybin, 15, 68, 73
psychedelics, 15–18, 22, 67, 68, 73–74; *see also* specific drugs
psychiatry, psychoanalysis, and views of, 3–4, 21, 45–46, 173–175; *see also* psychology; psychotherapy
psychoactive drugs, 57–59, 60–61; *see also* specific drugs
psychobiology, development of, 15–17, 19–20, 22
psychological dependence, 63–64; contrasted to addiction
psychology, 3–4, 27–28, 173, 190, 194–196
psychopharmacology, development of, 15, 20–21
psychosis, 18, 20, 21, 35, 68, 155
psychotherapy, 4, 7, 19–20, 147, 153, 181, 183, 186
psychotomimetics. *See* psychedelics
punishment center of brain, 71

Quaalude, 68, 72

ramp effect, 130, 131
Reagan, Ronald, 64
rebound: drug, 62, 134; romantic, 134
receptors. *See* postsynaptic receptors
rejection, 3, 10; and hysteroid dysphoria, 166–167, 169
religious/mystical experience, 16–17, 67, 115
reverse tolerance, 62
rewards, 41–43

Index

romantic love: as activator of pleasure center, 48–49; attraction in, 90, 91–96, 101–102, 117, 123–127, 141–142; biochemical factors in, 4–5, 7–8, 31, 48–49, 89–93, 96–101, 119–120, 152; challenge as attraction in, 123–124; defined, 87–88; disinhibition effect, 73, 76; drug-induced effects compared, 69–70, 92, 96–97, 99, 106; historical/cultural background, 87–88, 190–193, 200; and hysteroid dysphoria, 149–150, 166; idealization in, 88, 91–93, 129, 141, 142; limerance, 151–153; lithium and, 158–159; peak experiences in, 115–116; psychological viewpoint on, 3–4, 190–192, 194–196; rebound in, 134; response to breakup of, 5, 161–166, 201; "sensation seekers" and, 76–78, 83, 119, 130; separation anxiety and, 108, 113–114, 143; sexism and, 190, 196–197; and tranquillizer system, 72; withdrawal effect and, 63–64, 135, 146; see also love

Sachar, Edward J., 19
Schacter, Stanley, and Schacter-Singer experiment, 25–28, 30, 69
schizophrenia, 18, 21, 82, 155; mania misdiagnosed as, 79
school phobias, 10, 11, 111, 147
sedatives, 13, 59, 68, 72–73; see also specific drugs
Seleucus, 151
"sensation seekers," 76–78, 83, 119, 130
Separation and Loss (Bowlby), 109
separation anxiety, 10–11, 64, 70, 103–110; in love and romance, 64, 70–71, 113–114, 143, 146–147
serotonin, 81
setting, as factor in romance, 132
sexual attraction and arousal, 23, 24, 28–30, 88, 91, 92, 123, 194–195

sexism, in romantic love, 190, 196–197
Singer, Jerome, and Schacter-Singer experiment, 25–28, 30, 69
slide effect, 133
smile, as measure of emotion, 27, 53, 55
sodium lactate, 112, 113
spinal cord, 32, 33
Stelazine, 13
stimulants, 13, 68–70; effects compared to romantic attraction, 92, 96–97, 99, 106; see also specific drugs
Stratonice, 151
"Styles of Loving" (Lee), 117
stress, 19–20, 25–30, 37; see also anxiety
sympathetic nervous system, 24, 25, 187
synapses, 34–35

Tennov, Dorothy, 151
Thorazine, 13
"thrill and adventure" types, 76; see also "sensation seekers"
tolerance: drug, 60–61; in love, 61–62, 131
tranquillizers, 36, 71–72, 156; naturally occurring in body, 37, 58; see also specific drugs
tuberculosis, 82

unrequited love, 143, 150–153

Valium, and postsynaptic receptors for, 36, 37, 58, 68, 71–72

Walster, Elaine Hatfield, 124
Weiss, Robert S., 144
withdrawal: drug, 59, 60, 62–63, 132, 135; in nondrug situations, 63–64, 135, 146
Wyatt, Richard, 177

X rays, 54, 55

Zuckerman, Marvin, 76